New Aspects of
Quantity Surveying Practice

Dedication

to Peg

New Aspects of Quantity Surveying Practice

A text for all construction professionals

Second edition

Duncan Cartlidge

ELSEVIER

AMSTERDAM • BOSTON • HEIDELBERG • LONDON • NEW YORK • OXFORD
PARIS • SAN DIEGO • SAN FRANCISCO • SINGAPORE • SYDNEY • TOKYO
Butterworth-Heinemann is an imprint of Elsevier

Butterworth-Heinemann is an imprint of Elsevier
Linacre House, Jordan Hill, Oxford OX2 8DP
30 Corporate Drive, Suite 400, Burlington, MA 01803

First edition 2002
Second edition 2006

Notice
No responsibility is assumed by the publisher for any injury and/or damage to persons or property as a matter of products liability, negligence or otherwise, or from any use or operation of any methods, instructions or ideas contained in the material herein. Because of rapid advances in the medical sciences, in particular, independent verification and drug dosages should be made

British Library Cataloguing in Publication Data
A catalogue record for this book is available from the British Library

Library of Congress Cataloging in Publication Data
A catalog record for this book is available from the Library of Congress

ISBN-13: 978-0-7506-6841-5
ISBN-10: 0-7506-6841-5

For information on all Butterworth-Heinemann publications
visit our web site at http://books.elsevier.com

Printed and bound in the UK

06 07 08 09 10 10 9 8 7 6 5 4 3 2 1

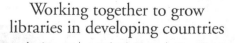

Contents

Foreword

Since the early 1990s there has been significant change in the construction industry, but the rate of that change will accelerate over the next decade as three factors have an increasing impact.

Firstly, the main focus both in the UK and around the world has been the need to find more effective methods of procurement. Discerning clients both in the private and public sector have identified the silo mentality and the fragmented nature of the design and construction process as very damaging characteristics that undermine the key objectives that lead to successful projects.

Much effort has been put into innovative procurement methods, the supply chain, partnering, strategic alliances, etc. These initiatives have led to major advances in the delivery of large and complex projects, but one of the major opportunities that has yet to be exploited to any great degree is the use of technology to aid the two major challenges faced by the quantity surveyor. These challenges straddle the design/construction process and involve the matching of the quantity surveyor's ability to produce feasibility estimates, cost plans and quantified schedules with the construction activities that involve operations and consume resources. The technology that is currently being used to design and manufacture Boeing's latest aircraft is also starting to address the design, communication, process, manufacture and assembly issues that up until now have the inhibited construction industry.

Secondly, far greater emphasis is being placed on the consideration of the value of cost and time. Clients that previously would have shied away from the appointment of leading edge designers are seeking ways of adding value to their developments. The appointment of innovative designers is increasingly the norm as clients recognise that they will enhance the function and form of the solution. This is in turn creating opportunities for quantity surveyors who can interpret the very earliest images and create secure platforms from which projects can be advanced. The priority as ever is the balance of risk and reward.

Apart from information technology and the establishment of secure project/cost management platforms, the third catalyst for change will be the impact of globalisation whereby the sourcing of capacity, components and capability will be used to enhance the viability of construction projects. Again it will afford major opportunities for quantity surveyors who have the leadership skills, management capability, understanding of process and global reach such that they can ensure greater certainty in areas where previously clients were confronted by performance difficulties and adversarial problems.

Rob Smith
Senior Partner
Davis Langdon LLP
20 February 2006

Preface to the second edition

Wanted: quantity surveyors

Four years have passed since the first edition of *New Aspects of Quantity Surveying Practice*. At that time *Building*, the well-known construction industry weekly, described quantity surveying as 'a profession on the brink' whilst simultaneously forecasting the imminent demise of the quantity surveyor and references to 'Ethel the Aardvark goes Quantity Surveying', had everyone rolling in the aisles. In a brave new world where confrontation was a thing of the past and where the RICS tried to deny quantity surveyor's existed at all, clearly there was no need of the profession! But wait. What a difference a few years can make for on 29 October 2004 the same publication that forecast the end of the quantity surveyor had to eat humble pie when the *Building* editorial announced that 'what quantity surveyors have to offer is the height of fashion – Ethel is history'. It would seem as if this came as a surprise to everyone, except quantity surveyors!

Ironically, in 2006 quantity surveyors are facing a very different challenge to the ones that were predicted in the late 1990s. Far from being faced with extinction, the problem now is a shortage of quantity surveyors that has reached crisis point, particularly in major cities like London. The 'mother of all recessions between 1990–1995' referred to in Chapter 1 had the effect of driving many professionals, including quantity surveyors out of the industry for good, as well as discouraging school leavers thinking of embarking on surveying degree courses. As a consequence there now is a generation gap in the profession and with the 2012 London Olympics on the horizon, as well as buoyant demand in most property sectors, many organisations are offering incentives and high salaries to attract and retain quantity surveying staff. In today's market-place a '30-something' quantity surveyor with 10 to 15 years experience is indeed a rare, but not endangered, species. It would also seem as though the RICS has had second thoughts about the future of the quantity surveyor. In 2006, the RICS announced that, after years of protest, the title quantity surveyor was to reappear as the Quantity Surveying and Construction Faculty.

A survey carried out by the Royal Bank of Scotland in 2005 indicated that quantity surveyors are the best paid graduate professionals. The new millennium found the construction industry and quantity surveying on the verge of a brave new world – an electronic revolution was coming, with wild predictions on the impact that IT systems and electronic commerce would have on the construction industry and quantity surveying practice. The reality is discussed in Chapter 5.

For the quantity surveyor, the challenges keep on coming. For many years the UK construction industry has flirted with issues such as whole life costs and sustainability/green issues . It now appears that these topics are being taken more seriously and are discussed in Chapter 3. The RICS Commission on Sustainable Development and Construction recently developed the following mission statement; 'To ensure that sustainability becomes and remains a priority issue throughout the profession and RICS' and committed itself to raising the profile of sustainability through education at all levels from undergraduate courses to the APC. In the public sector, the new Consolidated EU Public Procurement Directive due for implementation in 2006 now makes sustainability a criteria for contract awards and a whole raft of legislation due in the Spring 2006 has put green issues at the top of the agenda. Links have now been proved between the market value of a building and its green features and related performance.

Following the accounting scandals of the Enron Corporation in 2003, quantity surveyors are being called on to bring back accountability both to the public and private sectors and worldwide expansion of the profession continues with further consolidation and the emergence of large firms moving towards supplying broad business solutions tailored to particular clients and sectors of the market.

Where to next?

Duncan Cartlidge
www.duncancartlidge.co.uk

Preface to the first edition

The Royal Institution of Chartered Surveyors' Quantity Surveying Think Tank: *Questioning the Future of the Profession*, heard evidence that many within the construction industry thought Chartered Quantity Surveyors were: arrogant, friendless and uncooperative. In addition, they were perceived to add nothing to the construction process, failed to offer services which clients expected as standard and too few had the courage to challenge established thinking. In the same year, Sir John Egan called the whole future of quantity surveying into question in the Construction Industry Task Force report Rethinking Construction and if this weren't enough, a report by the University of Coventry entitled *Construction Supply Chain Skills Project* concluded that quantity surveyors are 'arrogant and lacking in interpersonal skills'. Little wonder then that the question was asked 'Will we soon be drying a tear over a grave marked "RIP Quantity Surveying, 1792–2000?"' Certainly the changes that have taken place in the construction industry during the past 20 years would have tested the endurance of the most hardy of beasts. Fortunately, the quantity surveyor is a tough and adaptable creature and to quote and paraphrase Mark Twain 'reports of the quantity surveyors death are an exaggeration'.

I have spent the past 30 years or so as a quantity surveyor in private practice, both in the UK and Europe, as well as periods as a lecturer in higher education. During this time I have witnessed a profession in a relentless search for an identity, from quantity surveyor to building economist, to construction economist, to construction cost advisor, to construction consultant, etc. I have also witnessed and been proud to be a member of a profession that has always risen to a challenge and has been capable of reinventing itself and leading from the front, whenever the need arose. The first part of the twenty-first century holds many challenges for the UK construction industry as well as the quantity surveyor, but of all the professions concerned with the procurement of built assets, quantity surveying is the one that has the ability and skill to respond to these challenges.

This book, therefore is dedicated to the process of transforming the popular perception that, in the cause of self preservation, the quantity surveyor is wedded to a policy of advocating aggressive price-led tendering with all the problems that this brings, to one of a professional who can help deliver high value capital projects on time and to budget with guaranteed life cycle costs. In addition it is hoped that this book will demonstrate beyond any doubt that the quantity surveyor is alive and well, adapting to the demands of construction clients and what's more, looking forward to a long and productive future. Nevertheless, there is still a long hill to climb. During the production of this book I have heard major construction clients call the construction industry 'very unprofessional' and the role of the quantity surveyor compared to that of a 'post box'.

In an address to the Royal Institution of Chartered Surveyors in November 2001, the same Sir John Egan, that had called the future of the quantity surveyor into question, but now as Chairman of the Egan Strategic Forum for Construction, suggested that the future for Chartered Surveyors in construction was to become process integrators, involving themselves in the process management of construction projects and that those who clung to traditional working practices faced an uncertain future. The author would wholeheartedly agree with these sentiments.

The quantity surveyor is dead – long live the quantity surveyor – masters of the process!

Duncan Cartlidge
www.duncancartlidge.co.uk

Acknowledgements

My thanks go to the following who have contributed to this book:

Rob Smith is Senior Partner at Davis Langdon LLP.

Graham Castle FRICS teaches at the Scott Sutherland School,
Robert Gordon University, Aberdeen. His research and consultancy interests include the role of information technology in the
built environment. He has extensive industrial experience in private practice, central government and local government as a
quantity surveyor, project manager and facilities manager.

Dr Richard Laing MRICS is Reader at the Scott Sutherland
School at RGU, Aberdeen. He has led a range of research com-
missions including Greenspace, Streetscapes, Urban Connections
and current work for NHS Estates. He has extensive experience
of research concerning holistic value assessment in the built environment, including studies on design evaluation, building conservation and innovative housing.

Mohammed Khirsk and Douglas I Gordon for their contribution
to Chapter 3.

John Goodall of FIEC, Brussels.

Crown copyright material is produced with the permission of the
Controller of HMSO and Queen's Printer for Scotland.

List of figures and tables

Figures

Tables

Abbreviations

2D	Two-dimensional
3D	Three-dimensional
CAD	Computer-aided design
CALS	Computer acquisition and lifetime support
CIC	Computer integrated construction
CPV	Common procurement vocabulary
CSF	Critical success factors
DXF	Data exchange format
EDI	Electronic data interchange
GPA	General procurement agreement
HTML	Hyper-text mark-up language
IAI	International Alliance for Interoperability
IFC	Industry foundation classes
IGES	Initial graphics exchange specification
IS	Information system
IT	Information technology
NAO	National Audit Office
OOP	Object-oriented programming
PDES	Product data exchange specification
PDM	Product data modeling
PFI	Private finance initiative
PPP	Public private partnerships
PPPP	Public private partnership programme
PSC	Public sector comparator
RICS	Royal Institution of Chartered Surveyors
SGML	Standard generalised mark-up language
SMEs	Small to medium-sized enterprises
SPC	Special purpose company
SPV	Special purpose vehicle
SQL	Structured query language

STEP   Standards for the exchange of product data
SWOT   Strengths weaknesses opportunities threats analysis
VRML   Virtual reality mark-up language
XML   Extensible mark-up language

1
The catalyst of change

Introduction

This chapter examines the root causes of the changes that took place in the United Kingdom construction industry and quantity surveying practice during the latter half of the twentieth century. It sets the scene for the remaining chapters, which go on to describe how quantity surveyors are adapting to new and emerging markets and responding to client-led demands for added value.

The catalyst of change

The construction industry is no stranger to fluctuations in workload; however, the period between 1990 and 1995 will be remembered, as an eminent politician once remarked, as 'the mother of all recessions'. Certainly, from the perspective of the UK construction industry, this recessionary phase was the catalyst for many of the changes in working practices and attitudes that have been inherited by those who survived this period and continue to work in the industry. As described in the following chapters, some of the pressures for change in the UK construction industry and its professions – including quantity surveying – have their origins in history, while others are the product of the rapid transformation in business practices that took place during the last decades of the twentieth century and still continue today. This book will therefore examine the background and causes of these changes, and then continue to analyse the consequences and effects on contemporary surveying practice.

Figure 1.1 The heady brew of change.

Historical overview

1990 was a watershed for the UK construction industry and its associated professions.

As illustrated in Figure 1.1, by 1990 a 'heady brew of change' was
being concocted on fires fuelled by a recession that was starting to
have an impact on the UK construction industry. The main ingre-
dients of this brew, in no particular order, were:

- The traditional UK hierarchical structure that manifested itself
  in a litigious, fragmented industry, where contractors and subcontractors were excluded from most of the design decisions.
- Changing patterns of workload due to the introduction of fee
  competition and compulsory competitive tendering.
- Widespread client dissatisfaction with the finished product.
- The emergence of privatisation and public private partnerships.

- The pervasive growth of information technology.
- The globalisation of markets and clients.

A fragmented and litigious industry

Boom and bust in the UK construction industry has been and will continue to be a fact of life (see Figure 1.2), and much of the industry, including quantity surveyors, had learned to survive and prosper quite successfully in this climate.

The rules were simple: in the good times a quantity surveyor earned fee income as set out in the Royal Institution of Chartered Surveyors' Scale of Fees for the preparation of, say, Bills of Quantities, and then in the lean times endless months or even years would be devoted to performing countless tedious re-measurements of the same work – once more for a fee. Contractors and subcontractors won work, albeit with very small profit margins, during the good times, and then when work was less plentiful they would turn their attentions to the business of the preparation of claims for extra payments for the inevitable delays and disruptions to the works. The standard forms of contract used by the industry, although heavily criticised by many, provided the impetus (if impetus were needed) to continue operating in this way. Everyone, including the majority of clients, appeared to be quite happy with the system, although in practice the UK construction industry was in many ways letting its clients down by producing buildings and other projects that were, in a high percentage of cases, over budget,

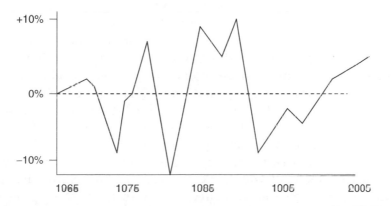

Figure 1.2 Construction output – percentage change 1965–2005 (Source: dti).

over time and littered with defects. Time was running out on this system, and by 1990 the hands of the clock were at five minutes before midnight. A survey conducted in the mid-1990s by *Property Week*, a leading property magazine, among private sector clients who regularly commissioned new buildings or refurbished existing properties, provided a snapshot of the UK construction industry at that time. In response to the question 'Do projects finish on budget?', 30 per cent of those questioned replied that it was quite usual for projects to exceed the original budget.

In response to the question 'Do projects finish on time?', once more over 30 per cent of those questioned replied that it was common for projects to overrun their planned completion by 1 or 2 months. Parallels between the construction industry ethos at the time of this survey and the UK car industry of the 1960s make an interesting comparison. Austin, Morris, Jaguar, Rolls Royce, Lotus and marques such as the Mini and MG were all household names during the 1960s. Today they are all either owned by foreign companies or out of business. At the time of writing the first edition of this book Rover/MG, then owned by Phoenix (UK), was the only remaining UK-owned carmaker, now Rover/MG has been confined to the scrap heap when it ceased trading in 2005 amidst bitter recriminations. Rover's decline from being the UK's largest carmaker in the 1960s is a living demonstration of how a country's leading industry can deteriorate, as well as being a stark lesson to the UK construction industry. The reasons behind the collapse of car manufacturing were flawed design, wrong market positioning, unreliability and poor build quality but importantly to this can be added; lack of investment in new technology, and a failure to move with the times and produce what the market, i.e. the end users, demanded. Therefore, when the first Datsun cars began to arrive from Japan in the 1970s and were an immediate success, it was no surprise to anyone except the UK car industry. The British car buyer, after overcoming initial reservations about purchasing a foreign car, discovered a product that had nearly 100 per cent reliability, contained many features as standard that were extras on British-built cars, were delivered on time, and benefited from long warranties. Instead of producing what they perceived to be the requirements of the British car buyer, Datsun had researched and listened to the needs of the market, seen the failings of the home manufacturers, and then produced a car to meet them. Not only had the Japanese car industry researched the market fully; it had also invested in plant and machinery to increase build quality and reduce defects in their cars. In addition, the entire manufacturing process was analysed and a lean supply chain established to ensure the maximum economies of production. The scale of the improvements achieved in the car industry are impressive, with the time from completed design to launch reduced from 40 to 15 months, and the supplier defects to five parts per million. So why by 1990 was the UK construction industry staring into the same abyss that the carmakers had faced 30 years earlier? In order to appreciate the situation that existed in the UK construction industry in the pre-1990 period it is necessary to examine the working practices of the UK construction industry, including the role of the contractor and the professions at this time. First, we will take a look back at recent history, and in particular at the events that took place in Europe in the first part of the nineteenth century and helped to shape UK practice (Goodall, 2000).

The UK construction industry – a brief history

Prior to the Napoleonic Wars, Britain, in common with its continental neighbours, had a construction industry based on separate trades. This system still exists in France as 'lots sépare', and variations of it can be found throughout Europe, including Germany. The system works like this: instead of the multi–traded main contractor that operates in the UK, each trade is tendered for and subsequently engaged separately under the co-ordination of a project manager, or 'pilote'. In France smaller contractors usually specialise in one or two trades, and it is not uncommon to find a long list of contractors on the site board of a construction project.

The Napoleonic Wars, however, brought change and nowhere more so than in Britain – the only large European state that Napoleon failed to cross or occupy. Paradoxically, the lasting effect the Napoleonic Wars had on the British construction industry was more profound than on any other national construction industry in Europe.

Whilst it is true that no military action actually took place on British soil, nonetheless the government of the day was obliged to construct barracks to house the huge garrisons of soldiers that were then being transported across the English Channel. As the need for the army barracks was so urgent and the time to prepare drawings, specifications, etc. was so short, the contracts were let on a 'settlement by fair valuation based on measurement after completion of the works'. This meant that constructors were given the opportunity and encouragement to innovate and to problem solve – something that was progressively withdrawn from them in the years to come. The same need for haste, coupled with the sheer magnitude of the individual projects, led to many contracts being let to a single builder or group of tradesmen 'contracting in gross', and the general contractor was born. When peace was made the Office of Works and Public Buildings, which had been increasingly concerned with the high cost of measurement and fair value procurement, in particular in the construction of Buckingham Palace and Windsor Castle, decided enough was enough. In 1828, separate trades contracting was discontinued for public works in England in favour of contracting in gross. The following years saw contracting in gross (general contracting) rise to dominate, and with this development the role of the builder as an innovator, problem-solver and design team member was stifled to the point where contractors operating in the UK system were reduced to simple executors of the works and instructions (although in Scotland the separate trades system survived until the early 1970s). However, history had another twist, for in 1834 architects decided that they wished to divorce themselves from surveyors and establish the Royal Institute of British Architects (RIBA), exclusively for architects. The grounds for this great schism were that architects wished to distance themselves from surveyors and their perceived 'obnoxious commercial interest' in construction. The top-down system that characterises so much of British society was stamped on the construction industry. As with the death of separate trades contracting, the establishment of the RIBA ensured that the UK contractor was once again discouraged from using innovation. The events of 1834 were also responsible for the birth of another UK phenomenon, the quantity surveyor, and for another unique feature of the UK construction industry – post construction liability.

The ability of a contractor to re-engineer a scheme design in order to produce maximum buildability is a great competitive advantage, particularly on the international scene (see Table 1.1). As discussed in Chapter 2, a system of project insurance that is already widely available on the Continent is starting to make an appearance in the UK. Adopting this, the design and execution teams can safely circumvent their professional indemnity insurance and operate as partners under the protective umbrella of a single policy of insurance, thereby allowing the interface of designers and contractors. However, back to history. For the next 150 or so years the UK construction industry continued to develop along the lines outlined above, and consequently by the third quarter of the twentieth century the industry was characterised by powerful professions carrying out work on comparatively generous fee scales, contractors devoid of the capability to analyse and refine design solutions, forms of contract that made the industry one of the most litigious in Europe, and procurement systems based upon competition and selection by lowest price and not value for money. Some within the industry had serious concerns about procurement routes and documentation, the forms of contract in use leading to excess costs, suboptimal building quality and time delays, and the adversarial and conflict-ridden relationships between the various parties. A series of government-sponsored reports (Simon, 1944; Emmerson, 1962; Banwell, 1964) attempted to stimulate debate about construction industry practice, but with little effect.

It was not just the UK construction industry that was obsessed with navel-gazing during the last quarter of the twentieth century; quantity surveyors had also been busy penning numerous reports into the future prospects for their profession. The most notable of which were: The Future Role of the Chartered Quantity Surveyor (1983), Quantity Surveying 2000 – The Future Role of the Chartered Quantity Surveyor (1991) and the Challenge for Change: QS Think Tank (1998), all produced either directly by, or on behalf of, The Royal Institution of Chartered Surveyors. The 1971 report, The Future Role of the Quantity Surveyor (RICS), was the product of a questionnaire sent to all firms in private practice together with a limited number of public sector organisations; sadly, but typically, the survey resulted in a mere 35 per cent response rate. The report paints a picture of a world where the quantity surveyor was primarily a producer of Bills of Quantities; indeed, the report comes to the conclusion that the distinct competence of the quantity surveyor of the 1970s was measurement – a view, it should be added, still shared by many today. In addition, competitive single stage tendering was the norm, as was the practice of receiving most work via the patronage of an architect. It was a profession where design and construct projects were rare, and quantity surveyors were discouraged from forming multidisciplinary practices and encouraged to adhere to the scale of fees charges. The report observes that clients were becoming more informed, but there was little advice about how quantity surveyors were to meet this challenge. A mere 25 years later the 1998 report, The Challenge for Change, was drafted in a business climate driven by information technology, where quantities generation is a low-cost activity and the client base is demanding that surveyors demonstrate added value. In particular, medium-sized quantity surveying firms (i.e. between 10 and 250 employees) were singled out by this latest report to be under particular pressure owing to:

- Competing with large practices' multiple disciplines and greater specialist knowledge base
- Attracting and retaining a high quality workforce
- Achieving a return on the necessary investment in IT
- Competing with the small firms with low overheads.

Consequently, the surveying profession has been predicted to polarise into two groups: the large multidisciplinary practices capable of matching the problem-solving capabilities of the large accountancy-based consulting firms, and small practices that can offer a fast response from a low cost base for clients, as well as providing services to their big brother practices. Interestingly, The Challenge for Change report also predicts that the distinction between contracting and professional service organisations will blur, a quantum leap from the 1960s, when chartered surveyors were forced to resign from their institution if they worked for contracting organisations! The trend for mergers and acquisitions continues, although it has to be said not without its problems, with the largest quantity surveying firms developing into providers of broad business solutions.

The British system compared

The following studies give the opportunities to directly compare the British system of procurement and project management with that of a European neighbour: France. In the mid-1990s, Graham Winch and Andrew Edkins carried out a study on the construction of two identical buildings needed to house a security scanning system as part of the Eurotunnel project. A leading UK architectural practice was commissioned for the design on both sides of the channel, who in turn procured medium-sized British and French firms for the construction. The resultant projects gave a unique opportunity to compare project performance in the two countries with a functionally equivalent building, a common design and a single client. The final analysis demonstrated how the French performed much better than the British in terms of out turn costs and completion times, despite the fact that both project teams faced similar challenges, largely generated by problems with scanning technology, yet the French team coped with them more smoothly. Why was this?

The answer would seem to lie in the differences in the organisation of the two projects:

- The French contract included detail design, the norm in France. The British contractor was deemed not capable of entering into a design and build contract due to the requirements for design information under the JCT form of contract.
- The French contractor re-engineered the project, simplifying the design and taking out unnecessary costs. This was possible because of the single point project liability that operates in France.
- Under the French contract, the British architect could not object to these contractor-led changes. Under the JCT contract, professional indemnity considerations meant that the architect refused to allow the British contractor to copy the French changes.
- The simplified French design was easier, cheaper and quicker to build. This meant that there was room for manoeuvre as the client induced variations mounted, whereas the British run project could only cope by increasing programme and budget. Once the project began to run late, work on construction became even less effective as the team had to start working out of sequence around the installation of scanning equipment. The researcher's conclusion was that British procurement arrangements tend to generate complexity in project organisation, while the engineering capabilities of French contractors mean that they are able to simplify the design. Indeed, they argue that it is these capabilities that are essential to the French contractor's ability to win contracts.

A second comparison in approaches to construction design and procurement was published in 2004 by the Building Design Partnership, entitled 'Learning from French Hospital Design'. Given the massive hospital building programme in the UK, that is planned to continue until at least 2010, the study compared French hospitals with newly built UK hospitals not only from the point of view of design quality, but also value for money. The results of the study are given in Table 1.1.

Health warning! When interpreting the cost data in Table 1.1, it should be remembered that direct comparison of cross-border cost

Table 1.1 UK and French hospital costs compared

	Floor Area (m²)	Total Cost (€)	Building Cost (€/m²)
UK examples			
Macclesfield	3353	7 182 634	2142
Hillingdon	3600	5 495 717	1527
Warley	8940	17 853 354	2103
Halton, Runcorn	5493	7 698 900	1402
French examples			
Montreuil sur Mer	19 691	16 776 184	852
St. Chamond	6953	11 897 767	1171
Armberieu	10 551	9 202 711	872
Chateauroux	4994	6 726 183	1347

Source: Building Design Partnership 2004.

information is notoriously difficult due to a range of factors including building regulations and other statutory controls. Even so

- French hospitals cost between half and two thirds of UK hospitals per m², but per bed they are more or less similar. Area per bed, however, is much higher in France, with single bed wards used universally. The report therefore argues that French bed space outperforms its UK counterparts.
- Building service costs, i.e. mechanical and electrical installation, in France are less than half of those of the UK, with French comments that the UK overspecifies. More ambitious automation and ICT are also used in France.
- Contractor-led detail design seems to lie behind much of the economy of means; many Egan-advocated processes are used. Interestingly consultants' fees, compared to the UK are high as a percentage of cost.
- In spite of the fact that labour and material costs are higher in France than the UK, although concrete, France's main structural material is 75 per cent of the UK cost, out-turn costs over a range of building types, not simply hospitals are cheaper. However, data released by Gardiner and Theobald seems to indicate recent trends, due in part to the differentials between the British pound and the Euro, has seen the gap close.
- The design quality of French hospitals is generally high, while in the UK standards achieved recently have been disappointing and have come in for some criticism.

Compared with many European countries, UK construction produces high output costs to customers from low input costs of professional, trade labour and materials. This fact is at the root of the Egan critique, pointing out that the UK has a wasteful system which would cost even more if UK labour rates were equal to those found in Europe. The waste in the system, 10 years plus on from Latham, is still estimated to be around 30 per cent. Looking at French design and construction it is possible to see several of the Egan goals in place, but in ways specific to France. Whilst the design process begins with no contractor involvement, they become involved sooner than in the UK and take responsibility for much of the detailed design and specification. They are more likely to buy standard components and systems from regular suppliers with well developed supply chains, rather than on a project by project basis. Constructional simplicity follows from the French approach with French architects having little control of details and not appearing to worry too much about doors and window details for example. In the case of French hospitals despite the lower cost, the projects contain very sophisticated technology with ICT systems becoming very ambitious.

Therefore a simple cost comparison demonstrates that French hospital out turn costs are cheaper than in the UK, but what of added value? Health outcomes in France are generally superior to those in the UK due to factors such as bed utilisation and patient recovery times and single rooms instead of multi-patient wards stop the spread of dangerous, so-called super bugs, such as MRSA.

Keeping the focus on Europe, for many observers the question of single-point or project liability – the norm in many countries, such as Belgium and France – is pivotal in the search for adding value to the UK construction product, and is at the heart of the other construction industries' abilities to re-engineer designs. Single-point project liability insurance is insurance that protects all the parties involved in both the design and the construction process against failures in both design and construction of the works for the duration of the policy. The present system, where some team members are insured and some not, results in a tendency to design defensively, caveat all statements and advice with exclusions of liability, and seek help from no other members of the team – not a recipe for teamwork. In the case of a construction management contract, the present approach to latent defect liability can result in the issue of 20–30 collateral warranties, which facilitates the creation of a contractual relationship where one would otherwise not exist in order that the wronged party is then able to sue under contract rather than rely on the tort of negligence. Therefore in order to give contractors the power truly to innovate and to use techniques like value engineering (see Chapter 2), there has to be a fundamental change in the approach to liability. Contract forms could be amended to allow the contractor to modify the technical design prior to construction, with the consulting architects and engineers waiving their rights to interfere.

If this approach is an option, then why does the UK construction industry still fail to produce the goods? The new Wembley Stadium and Pickett's Lock, both proposed venues for the World Athletics Championship in 2005, are prime examples of the traditional UK approach – namely, this is the design, this is the cost, that's it, we've had it! The principal problems behind the failure of these two high-profile fiascos were no business case, little or no understanding of the needs of the client, and the inability of a contractor to re-engineer the proposals and produce alternatives. The result – grandiose designs with large price tags and a complete disregard of the need to pay back the cost of the project from revenues generated by the built asset, in this case a sports stadium. By 2006 it appears as though the new Wembley Stadium will be delivered albeit 2 months late and £106 million over budget and not without the customary nail biting finish.

Opponents of the proposal to introduce single-point liability cite additional costs as a negative factor. However, indicative costs given by Royal & SunAlliance seem to prove that these are minimal – for example, traditional structural and weatherproofing: 0.65–1.00 per cent of contract value total cover, including structural, weatherproofing, non-structural and mechanical and electrical; 1–2 per cent of contract value to cover latent defects for periods of up to 12 years, to tie in with the limitations provisions of contracts under Seal. As in the French system, technical auditors can be appointed to minimise risk and, some may argue, add value through an independent overview of the project.

Changing patterns of workload

The patterns of workload that quantity surveyors had become familiar with were also due to change. The change came chiefly from two sources:

1.   Fee competition and compulsory competitive tendering (CCT)
2.   The emergence of a new type of construction client.

Fee competition and compulsory competitive tendering

Until the early 1970s, fee competition between professional practices was almost unheard of. All the professional bodies published scales of fees, and competition was vigorously discouraged on the basis that a client engaging an architect, engineer or surveyor should base his or her judgement on the type of service and not on the level of fees. Consequently, all professionals within a specific discipline quoted the same fee. However, things were to change with the election of the Conservative Government in 1979. The new government introduced fee competition into the public sector by way of its compulsory competitive tendering programme (CCT), and for the first time professional practices had to compete for work in the same manner as contractors or subcontractors – i.e. they would be selected by competition, mainly on the basis of price. The usual procedure was to submit a bid based upon scale of fees minus a percentage. Initially these percentage reductions were a token 5 or 10 per cent, but as work became difficult to find in the early 1980s, practices offered 30 or even 40 per cent reduction on fee scales. It has been suggested that during the 1980s fee income from some of the more traditional quantity surveying services was cut by 60 per cent. Once introduced there was no going back, and soon the private sector began to demand the same reduction in fee scales. Within a few years the cosy status quo that had existed and enabled private practices to prosper had gone. The Monopolies and Mergers Commission's 1977 report into scales of fees for survey-ors' services led the Royal Institution of Chartered Surveyors to revise its byelaws in 1983 to reduce the influence of fee scales to the level of 'providing guidance' – the gravy train had hit the buffers!

Byelaw 24 was altered from:

> No member shall with the object of securing instructions or sup-planting another member of the surveying profession, knowingly attempt to compete on the basis of fees and commissions

to

> ... no member shall ... quote a fee for professional services with-out having received information to enable the member to assess the nature and scope of the services required.

With the introduction of fee competition, the average fee for quantity surveying services (expressed as a percentage of construction cost) over a range of new build projects was just 1.7 per cent! As a result, professional practices found it increasingly difficult to offer the same range of services and manning levels on such a reduced fee income. They had radically to alter the way they operated, or go out of business. However, help was at hand for the hard-pressed practitioner. The difficulties of trying to manage a practice on reduced fee scale income during the later part of the 1980s were mitigated by a property boom, which was triggered in part by a series of government-engineered events that combined to unleash a feeding frenzy of property development. In 1988, construction orders peaked at £26.3 billion, and the flames under the heady brew of change were dampened down, albeit only for a few years. The most notable of these events were:

- The so-called Stock Exchange 'Big Bang' of 1986, which had the direct effect of stimulating the demand for high-tech offices
- The deregulation of money markets in the early 1980s, which allowed UK banks for the first time to transfer money freely out of the country, and foreign finance houses and banks to lend freely on the UK market and invest in UK real estate
- The announcement by the Chancellor of the Exchequer, Nigel Lawson, of the abolition of double tax mortgage relief for domestic dwellings in 1987, which triggered an unprecedented demand for residential accommodation; the result was a massive increase in lending to finance this sector, as well as spiralling prices and land values
- Last but by no means least, the relaxation of planning controls, which left the way open for the development of out-of-town shopping centres and business parks.

However, most property development requires credit, and the boom in development during the late 1980s could not have taken place without financial backing. By the time the hard landing came in 1990, many high street banks with a reputation for prudence found themselves dangerously exposed to high-risk real estate projects. During the late 1980s, virtually overnight the banks changed from conservative risk managers to target-driven loan sellers, and by 1990 they found themselves with a total property-related debt of £500 billion. The phenomenon was not just confined to the UK. In France, for example, one bank alone, Credit Lyonnais, was left with

€ 10 billion of unsecured loss after property deals on which the bank had lent money collapsed because of oversupply and a lack in demand; only a piece of creative accountancy and state intervention saved the French bank from insolvency. The property market crash in the early 1990s occurred mainly because investors suffered a lack of confidence in the ability of real estate to provide a good return on investment in the short to medium term in the light of high interest rates, even higher mortgage rates, and an inflation rate that doubled within 2 years. In part it was also brought about by greed because of the knowledge that property values had historically seldom delivered negative values.

The emergence of a new type of construction client

Another vital ingredient in the heady brew of change was the emergence of a new type of construction client. Building and civil engineering works have traditionally been commissioned by either public or private sector clients. The public sector has been a large and important client for the UK construction industry and its professions. Most government bodies and public authorities would compile lists or 'panels' of approved quantity surveyors and contractors for the construction of hospitals, roads and bridges, social housing, etc., and inclusion on these panels ensured that they received a constant and reliable stream of work. However, during the 1980s the divide between public and private sectors was to blur. The Conservative Government of 1979 embarked upon an energetic and extensive campaign of the privatisation of the public sector that culminated in the introduction of the Private Finance Initiative in 1992 (see Chapter 4). Within a comparatively short period there was a shift from a system dominated by the public sector to one where the private sector was growing in importance. Despite this shift to the private sector the public sector still remains influential; in 2005, for example, it accounted for 37 per cent or £25 billion of the UK civil engineering and construction industry's business, with a government pledge to maintain this level of expenditure. Nevertheless, the privatisation of the traditional public sector resulted in the emergence of major private sector clients such as the British Airports Authority, privatised in 1987, with an appetite for change and innovation. This new breed of client was, as the RICS had predicted in its 1971 report on the future of quantity surveying, becoming more knowledgeable about the construction process, and such clients were not prepared to sit on their hands while the UK construction industry continued to under perform. Clients such as Sir John Egan, who in July 2001 was appointed Chairman of the Strategic Forum for Construction, became major players in the drive for value for money. The poor performance of the construction industry in the private sector has already been examined, however, if anything, performance in the public sector paints an even more depressing picture. This performance was scrutinised by the National Audit Office (NAO) in 2001 in its report Modernising Construction (Auditor General, 2000), which found that the vast majority of projects were over budget and delivered late. So dire has been the experience of some public sector clients – for example, the Ministry of Defence – that new client-driven initiatives for procurement, have been introduced. In particular there were a number of high-profile public projects disasters such as the new Scottish Parliament in Edinburgh, let on a management contracting basis which rose in cost from approximately £100 million to £450 million and was delivered in 2004 – 2 years late and with a total disregard for life cycle costs.

If supply chain communications were polarised and fragmented in the private sector, then those in the public sector were even more so. A series of high-profile cases in the 1970s, in which influential public officials were found to have been guilty of awarding construction contracts to a favoured few in return for bribes, instilled paranoia in the public sector, which led to it distancing itself from contractors, subcontractors and suppliers – in effect from the whole supply chain. At the extreme end of the spectrum this manifested itself in public sector professionals refusing to accept even a diary, calendar or a modest drink from a contractor in case it was interpreted as an inducement to show bias. In the cause of appearing to be fair, impartial and prudent with public funds, most public contracts were awarded as a result of competition between a long list of contractors on the basis of the lowest price. The 2001 National Audit Office report suggests that the emphasis on selecting the lowest price is a significant contributory factor to the tendency towards adversarial relationships. Attempting to win contracts under the 'lowest price wins' mentality leads firms to price work unrealistically low and then seek to recoup their profit margins through contract variations arising from, for example design changes and other claims leading to disputes and litigation. In an attempt to eradicate inefficiencies, the public sector commissioned a number of studies such as The

Levene Efficiency Scrutiny in 1995, which recommended that departments in the public sector should:

- Communicate better with contractors to reduce conflict and disputes
- Increase the training that their staff receive in procurement and risk management
- Establish a single point for the construction industry to resolve problems common to a number of departments. The lack of such a management tool was identified as one of the primary contributors to problems with the British Library project.

In June 1997 it was announced that Compulsory Competitive Tendering would be replaced with a system of Best Value in order to introduce, in the words of the local government minister Hilary Armstrong, 'an efficient, imaginative and realistic system of public sector procurement'. Legislation was passed in 1999, and from 1 April 2000 it became the statutory duty of the public sector to obtain best value. Best value will be discussed in more detail in Chapter 4.

In 2002 the Office of Government Commerce announced that the preferred methods of procurement for the UK public sector are to be:

- Public private partnerships
- Prime contracting
- Design and build.

The information technology revolution

As measurers and information managers, quantity surveyors have been greatly affected by the information technology revolution. Substantial parts of the chapters that follow are devoted to the influence that IT has had and will continue to have, both directly and indirectly, on the quantity surveying profession. However, this opening chapter would not be complete without a brief mention of the contribution of IT to the heady brew of change. To date, mainly individual IT packages have been used or adapted for use by the quantity surveyor – for example, spreadsheets. However, the next few years will see the development of IT packages designed specifically for tasks such as measurement and quantification, which will fundamentally change working practices. The speed of development has been breathtaking. In 1981 the Department of the Environment developed and used a computer-aided bill of quantities production package called 'Enviro'. This then state-of-the-art system required the quantity surveyor to code each measured item, and on completion the codes were sent to Hastings, on the south coast of England, where a team of operators would input the codes, with varying degrees of accuracy, into a mainframe computer. After the return of the draft bill of quantities to the measurer for checking, the final document was then printed, which in most cases was 4 weeks after the last dimensions were taken off!

In recent years architects have made increasing use of computer-aided design (CAD) in the form of 2D drafting and 3D modelling for the production of project information. A recent report by the Construction Industry Computing Association and entitled Architectural IT Usage and Training Requirements indicated that in architectural practices with more than six staff, between 95 and 100 per cent of all those questioned used 2D drafting to produce information. This shift from hand drawn drafting to IT-based systems has allowed packages to be developed that link the production of drawings and other information to their measurement and quantification, thereby revolutionising the once labour-intensive bill of quantities preparation procedure. Added to this, the spread of the digital economy means that drawings and other project information can be produced, modified and transferred globally. One of the principal reasons for quantity surveyors' emergence as independent professionals during the Napoleonic Wars and their subsequent growth to hold a pivotal role in the construction process had, by the end of the 1990s, been reduced to a low-cost IT operation.

Those who mourn the demise of traditional methods of bill of quantities production should at least take heart that no longer will the senior partner be able to include those immortal lines in a speech at the annual Christmas office party – 'you know after 20 years of marriage my wife thinks that quantity surveying is all about taking off and working up' – pause for laughter!

As mentioned previously, there had been serious concern both in the industry and in government about the public image of UK Construction plc. The 1990 recession had opened the wounds in the construction industry and shown its vulnerability to market pressures. Between 1990 and 1992 over 3800 construction enterprises became insolvent, taking with them skills that would be badly needed in the future. The professions also suffered a similar haemorrhage of skills as the value of construction output fell by double digit figures year on year. The recession merely highlighted what had been apparent for years: the UK construction industry and its professional advisors had to change. The heady brew of change was now complete, but concerns over whether or not the patient re-alised the seriousness of the situation still gave grounds for con-cern. The message was clear: industry and quantity surveying must change or, like the dinosaur, be confined to history!

Response to change

In traditional manner, the UK construction industry turned to a re-port to try to solve its problems. In 1993 Sir Michael Latham, an academic and politician, was tasked to prepare yet another review, this time of the procurement and contractual arrangements in the United Kingdom construction industry. In July 1994, Constructing The Team (or The Latham Report, as it became known) was published. The aims of the initiative were to reduce conflict and lit-igation, as well as to improve the industry's productivity and com-petitiveness. The construction industry held its breath – was this just another Banwell or Simon to be confined, after a respectful pe-riod, to gather dust on the shelf? Thankfully not! The UK con-struction industry was at the time of publication in such a fragile state that the report could not be ignored. That's not to say that it was greeted with open arms by everyone – indeed, the preliminary report Trust and Money, produced in December 1993, provoked profound disagreement in the industry and allied professions.

Latham's report found that the industry required a good dose of medicine, which the author contended should be taken in its en-tirety if there was to be any hope of a revival in its fortunes. The Latham Report highlighted the following areas as requiring partic-ular attention to assist UK construction industries to become and be seen as internationally competitive:

- Better performance and productivity, to be achieved by using adjudication as the normal method of dispute resolution, the adoption of a modern contract, better training, better tender evaluation, and the revision of post-construction liabilities to be more in line with, say, France or Spain, where all parties and not just the architect are considered to be competent players and all of them therefore are liable for non-performance for up to 10 years

- The establishment of well-managed and efficient supply chains and partnering agreements
- Standardisation of design and components, and the integration of design, fabrication and assembly to achieve better buildability and functionality
- The development of transparent systems to measure performance and productivity both within an organisation and with competitors
- Teamwork and a belief that every member of the construction team from client to subcontractors should work together to produce a product of which everyone can be justifiably proud.

The Latham Report placed much of the responsibility for change on clients in both public and private sectors. For the construction industry, Latham set the target of a 30 per cent real cost reduction by year 2000, a figure based on the CRINE (Cost Reduction Initiative for the New Era) review carried out in the oil and gas industries a few years previously (CRINE, 1994). The CRINE review was instigated in 1992, with the direct purpose of identifying methods by which to reduce the high costs in the North Sea oil and gas industry. It involved a group of operators and contractors working together to investigate the cause for such high costs in the industry, and also to produce recommendations to aid the remedy of such. The leading aim of the initiative was to reduce development and production costs by 30 per cent, this being achieved through recommendations such as the use of standard equipment, simplifying and clarifying contract language, removing adversarial clauses, rationalisation of regulations, and the improvement of credibility and quality qualifications. It was recommended that the operators and contractors work more closely, pooling information and knowledge, to help drive down the increasing costs of hydrocarbon products and thus indirectly promote partnering and alliancing procurement strategies (see Chapter 3). The CRINE initiative recommendations were accepted by the oil and gas industries, and it is now widely accepted that without the use of partnering/alliancing a great number of new developments in the North Sea would not have been possible. Shell UK Exploration and Production reported that the performance of the partners in the North Fields Unit during the period 1991–1995 resulted in an increase in productivity of 25 per cent, a reduction in overall maintenance costs of 31 per cent in real terms, and a reduction in platform 'down time' of 24 per cent. Could these dramatic statistics be replicated in the construction industry? 'C' is not only for construction but also for conservative, and many sectors of the construction industry considered 30 per cent to be an unrealistically high and unreachable target. Nevertheless, certain influential sections of the industry, including Sir John Egan and BAA, accepted the challenge and went further declaring that 50 or even 60 per cent savings were achievable. It was the start of the client-led crusade for value for money.

The Latham Report spawned a number of task groups to investigate further the points raised in the main report, and in October 1997, as a direct result of one of these groups, Sir John Egan, a keen advocate of Sir Michael Latham's report and known to be a person convinced of the need for change within the industry, was appointed as head of the Construction Task Force. One of the task force's first actions was to visit the Nissan UK car plant in Sunderland to study the company's supply chain management techniques and to determine whether they could be utilised in construction (see Chapter 2). In June 1998 the task force published the report Rethinking Construction (DoE, 1998), which was seen as the blueprint for the modernisation of the systems used in the UK construction industry to procure work. As a starting point, Rethinking Construction revealed that in a survey of major UK property clients, many continued to be dissatisfied with both contractors' and consultants' performance. Added to the now familiar concerns about failure to keep within agreed budgets and completion schedules, clients revealed that:

- More than a third of them thought that consultants were lacking in providing a speedy and reliable service
- They felt they were not receiving good value for money insofar as construction projects did not meet their functional needs and had high whole-life costs
- They felt that design and construction should be integrated in order to deliver added value.

Frustrated by the lack of change in the construction industry, Egan's last act before moving on from the task force in 2002 was to pen his final report 'Accelerating Change'.

As for quantity surveyors, the 1990s ended with perhaps the unkindest cut of all. The RICS, in its Agenda for Change initiative, replaced its traditional divisions (which included the Quantity Surveying Division) with 16 faculties, not unlike the system operated by Organisme Professionel de Qualification Technique des

Economistes et Coordonnateurs de la Construction (OPQTECC), the body responsible for the regulation of the equivalent of the quantity surveyor in France. It seemed to some that the absence of a quantity surveying faculty would result in the marginalisation of the profession; however, the plan was implemented in 2000, with the Construction Faculty being identified as the new home for the quantity surveyor within the RICS. This move however was not taken lying down by the profession, disillusioned quantity surveyors threatened the RICS with legal action to reverse the decision, a new institute was formed by the James Knowles organisation, especially for quantity surveyors and the *Builder* group began in 2004 to publish a new weekly magazine for quantity surveyors: *QS News*. By 2006 it appeared that the RICS had a change of mind, with references to quantity surveyors reappearing on the RICS website and a restructuring of the faculties to include the return of the Quantity Surveying and Construction Faculty.

Beyond the rhetoric

How are the construction industry and the quantity surveyor rising to the challenges outlined in the previous pages?

When the much-respected quantity surveyors Arthur J. and Christopher J. Willis penned the foreword to the eighth edition of their famous book *Practice and Procedure for the Quantity Surveyor* in 1979, the world was a far less complicated place. Diversification into new fields for quantity surveyors included heavy engineering, coal mining and 'working abroad'. In the Willis book, the world of the quantity surveyor was portrayed as a mainly technical back office operation providing a limited range of services where, in the days before compulsory competitive tendering and fee competition, 'professional services were not sold like cans of beans in a supermarket'. The world of the Willis's was typically organised around the production of bills of quantities and final accounts, with professional offices being divided into pre- and post-contract services. This model was uniformly distributed across small and large practices, the main difference being that the larger practices would tend to get the larger contracts and the smaller practices the smaller contracts. This state of affairs had its advantages, as most qualified quantity surveyors could walk into practically any office and start work immediately; the main distinguishing feature between practices A and B was usually only slight differences in the format of taking-off paper. However, owing to the changes that have taken place not only within the profession and the construction industry, but also on the larger world stage (some of which have been outlined in this chapter), the world of the Willis's has, like the British Motor Car Industry, all but disappeared forever.

In the early part of the twenty-first century, the range of activities and sectors where the quantity surveyor is active is becoming more and more diverse. The small practice concentrating on traditional pre- and post-contract services is still alive and healthy. However, at the other end of the spectrum the larger practices are now rebadged as international consulting organisations and would be unrecognisable to the Willis's. The principal differences between these organisations and traditional large quantity surveying practices are generally accepted to be the elevation of client focus and business understanding and the move by quantity surveyors to develop clients' business strategies and deliver added value. As discussed in the following chapters, modern quantity surveying involves working in increasingly specialised and sectorial markets where skills are being developed in areas including strategic advice in the PFI, partnering, value and supply chain management.

From a client's perspective, it is not enough to claim that the quantity surveyor and/or the construction industry is delivering a better value service; this has to be demonstrated. Certainly there seems to be a move by the larger contractors away from the traditional low-profit, high-risk, confrontational procurement paths toward deals based on partnering and PFI and the team approach advocated by Latham. Table 1.2 illustrates the trend away from traditional lump sum contract based on bills of quantities.

Table 1.2 Trends in methods procurement – number of contracts

	1985	1987	1989	1991	1993	1995	1998	2001
Lump sum – firm BQ(%)	42.8	35.6	39.7	29.0	34.5	39.2	30.8	19.6
Lump sum (spec and drawings)(%)	47.1	55.4	49.7	59.2	45.6	43.7	43.9	62.9
Design and build(%)	3.6	3.6	5.2	9.1	16.0	11.8	20.7	13.9
Construction management(%)	–	–	0.2	0.2	0.4	1.3	0.8	0.5
Partnering(%)	–	–	–	–	–	–	–	0.6
Others(%)	6.5	5.4	5.2	2.5	3.5	4.0	3.8	2.5

Source: RICS Contracts in Use 2003.

The terms of reference for the Construction Industry Task Force concentrated on the need to improve construction efficiency and to establish best practice. The industry was urged to take a lead from other industries, such as car manufacturing, steel making, food retailing and offshore engineering, as examples of market sectors that had embraced the challenges of rising world-class standards and invested in and implemented lean production techniques. Rethinking Construction identified five driving forces that needed to be in place to secure improvement in construction and four processes that had to be significantly enhanced, and set seven quantified improvement targets, including annual reductions in construction costs and delivery times of 10% and reductions in building defects of 20% (see Table 1.3).

Table 1.3 Rethinking Construction recommendations

The five key drivers that need to be in place to achieve better construction are:

1. Committed leadership
2. Focus on the customer
3. Integration of process and team around the project
4. A quality-driven agenda
5. Commitment to people.

The four key projected processes needed to achieve change are:

1. Partnering the supply chain – development of long-term relationships based on continuous improvement with a supply chain
2. Components and parts – a sustained programme of improvement for the production and the delivery of components
3. Focus on the end product – integration and focusing of the construction process on meeting the needs of the end user
4. Construction process – the elimination of waste.

The seven annual targets capable of being achieved in improving the performance of construction projects are:

1. To reduce capital costs by 10%
2. To reduce construction time by 10%
3. To reduce defects by 20%
4. To reduce accidents by 20%
5. To increase the predictability of projected cost and time estimates by 10%
6. To increase productivity by 10%
7. To increase turnover and profits by 10%.

The report also drew attention to the lack of firm quantitative information with which to evaluate the success or otherwise of construction projects. Such information is essential for two purposes:

1. To demonstrate whether completed projects have achieved the planned improvements in performance
2. To set reliable targets and estimates for future projects based on past performances.

It has been argued that organisations like the Building Cost Information Service have been providing a benchmarking service for many years through its tender-based index. Additionally, what is now required is a transparent mechanism to enable clients to determine for themselves which professional practice, contractor, subcontractor, etc. delivers best value.

Benchmarking

Benchmarking is a generic management technique that is used to compare performance between varieties of strategically important performance criteria. These criteria can exist between different organisations or within a single organisation provided that the task being compared is a similar process. It is an external focus on internal activities, functions or operations aimed at achieving continuous improvement (Leibfried and McNair, 1994). Construction, because of the diversity of its products and processes is one of the last industries to embrace objective performance measurements. There is a consensus among industry experts that one of the principal barriers to promote improvement in construction projects is the lack of appropriate performance measurement and this is also referred to in Chapter 2 in relation to whole-life costs calculations. For continuous improvement to occur it is necessary to have performance measures which check and monitor performance, to verify changes and the effect of improvement actions, to understand the variability of the process and in general it is necessary to have objective information available in order to make effective decisions. Despite the late entry of benchmarking to construction, this does not diminish the potential benefits that could be derived. However, it gives some indication of the fact that there is still considerable work to be undertaken both to define the areas where benchmarking might be valuable and the methods of measurement. The current benchmarking and KPI programme in the UK construction industry has been headlined as a way to improve underperformance. However despite the production of several sets of KPIs, large scale improvement still remains as elusive as ever. Why is this?

The Xerox Corporation in America is considered to be the pioneer of benchmarking. In the late 1970s Xerox realised that it was on the verge of a crisis when Japanese companies were marketing photocopiers cheaper than it cost Xerox to manufacture a similar product. It is claimed that by benchmarking Xerox against Japanese companies it was able to improve their market position and the company has used the technique ever since to promote continuous improvement. Yet again another strong advocate of benchmarking is the automotive industry who successfully employed the technique to reduce manufacturing faults. Four types of benchmarking can broadly be defined: internal, competitive, functional and generic (Lema and Price, 1995). However, Carr and Winch (1998) whilst regarding these catagories as important suggest that a more useful distinction in terms of methodology is that of output and process benchmarking. Interestingly, Winch (1996) discovered that sometimes the results from a benchmarking exercise could be surprising, as illustrated in Table 1.4, which shows the performance of the Channel Tunnel project relative to other multi-million dollar mega infrastructure projects throughout the world using benchmarks established by the RAND Corporation. The results are surprising because the Channel Tunnel is regularly cited as an example of just how bad UK construction industry is at delivering prestige projects. By contrast, the Winch benchmarking exercise demonstrated that the Channel Tunnel project faired better than average when measured against a range of performance criterion.

Table 1.4 Performance benchmarking mega projects

Performance Criterion	Mega Projects Average	Channel Tunnel
Budget increase	88%	69%
Programme overrun	17%	14.2%
Conformance overrun	53% performance not up to expectations	As expected
Operational profitability	72% unprofitable	Operationally profitable but overwhelmed by finance charges

Source: Winch (1996).

Since the Winch study in 1996 there has been a tailing off of traffic using the Channel Tunnel and it is now obvious that projections of growth of users was grossly optimistic.

Measuring performance

Through the implementation of performance measures (what to measure) and selection of measuring tools (how to measure), an organisation or a market sector communicates to the outside world and clients the priorities, objectives and values that the organisation or market sector aspires to. Therefore the selection of appropriate measurement parameters and procedures is very important to the integrity of the system.

It is now important to distinguish between benchmarks and benchmarking. It is true to say that most organisations that participate in the production of Key Performance Indicators (KPIs) for the Construction Best Practice Programme (CBPP) has to produce benchmarks. Since the late 1990s, there has been a widespread government-backed campaign to introduce benchmarking into the construction industry with the use of so-called Key Performance Indicators. The objectives of the benchmarking as defined by the Office of Government Commerce are illustrated in Figure 1.3. Benchmarks provide an indication of position relative to what is considered optimum practice and hence indicate a goal to be obtained, but while useful for getting a general idea of areas requiring performance improvement, they provide no indication of the mechanisms by which increased performance may be brought about. Basically it tells us that we are underperforming but it does not give us the basis for the underperformance. The production of KPIs which has been the focus of construction industry initiatives to date therefore has been concentrated on output benchmarks. A much more beneficial approach to measurement is process benchmarking described by (Pickerell and Garnett, 1997) 'as analyzing why your current performance is what it is, by examining the process your business goes through in comparison to other organisations that are doing better and then implementing the improvements to boost performance'. The danger with the current enthusiasm in the construction industry for KPIs is that outputs will be measured and presented, but processes will not be improved as the underlining causes will not be understood. According to Carr and Winch (1998), many recent benchmarking

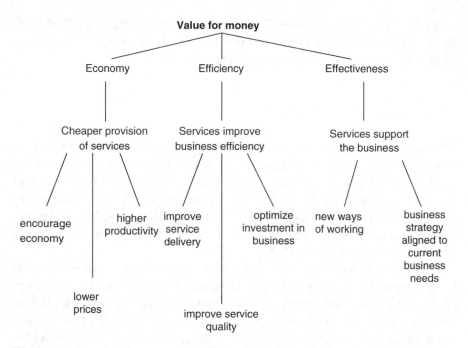

Figure 1.3 Value for money and benchmarking drivers (Source: OGC).

initiatives in the construction industry have shown that while the principles have been understood and there is much discussion about its potential 'no one is actually doing the real thing'. Benchmarking projects have tended to remain as strategic goals at the level of senior management.

The performance measures selected by the Construction Industry Best Practice Programme are:

Project performance
    Client satisfaction – product
    Client satisfaction – service
    Defects – product
    Predictability – cost
    Predictability – time
    Construction – cost
    Construction – time

Company performance
    Profitability
    Productivity

Safety
Respect for people
Environmental

and are measured at five key stages throughout the lifetime of a project. The measurement tools range from crude scoring on a 1 to 10 basis to the number of reportable accidents per 100 000 employees. For benchmarking purposes, the construction industry is broken down into sectors such as public housing and repair and maintenance.

The balanced scorecard

Robert Kaplan and David Norton at Harvard Business School developed a new approach to performance measurement in the early 1990s. They named the system the 'balanced scorecard' (BSC). Recognising some of the weaknesses and vagueness of previous management approaches, the balanced scorecard approach provides a clear prescription as to what organisations should measure in order to balance the financial perspective. The BSC is not just a measurement system, it is also a management system that enables organisations to clarify their vision and strategy and translate them into action. The BSC methodology builds on some key concepts of previous management ideas such as Total Quality Management (TQM), including customer-defined quality, continuous improvement, employee empowerment and primarily measurement-based management and feedback.

Kaplan and Norton describe the innovation of the balanced scorecard as follows:

The balanced scorecard retains traditional financial measures. Because financial measures tell the story of past events, an adequate story for industrial age companies for which investments in long-term capabilities and customer relationships were not critical for success. These financial measures are inadequate, however, for guiding and evaluating the journey that information age companies must make to create future value through investment in customers, suppliers, employees, processes, technology, and innovation.

Metrics

You can't improve what you can't measure. Often used interchangeably with measurements. However, it is helpful to separate these definitions. Metrics are the various parameters or ways of looking at a process that is to be measured. Metrics define *what* is to be measured. Some metrics are specialised, so they can't be directly benchmarked or interpreted outside a mission-specific business unit, other measures will be generic. It is true that defining metrics is time consuming and has to be done by managers in their respective mission units. But once they are defined, they won't change very often. Some of the metrics are generic across all units, such as cycle time, customer satisfaction, employee attitudes, etc. Also, software tools are available to assist in this task. Metrics must be developed based on the priorities of the strategic plan, which provides the key business drivers and criteria for metrics managers most desire to watch. Processes are then designed to collect information relevant to these metrics and reduce it to numerical form for storage, display and analysis. Decision-makers examine the outcomes of various measured processes and strategies and track the results to guide the company and provide feedback.

The value of metrics is in their ability to provide a factual basis for defining:

- Strategic feedback to show the present status of the organisation from many perspectives for decision-makers
- Diagnostic feedback into various processes to guide improvements on a continuous basis
- Trends in performance over time as the metrics are tracked
- Feedback around the measurement methods themselves, and which metrics should be tracked
- Quantitative inputs to forecasting methods and models for decision support systems.

The BSC approach can be applied equally to both the public and private sectors, although objectives in a public sector organisation will tend to differ, for example financial considerations in the public sector will seldom be the primary objective for business systems. The BSC suggests that an organisation is viewed from four perspectives, as illustrated in Figure 1.4 and to develop metrics, collect data and analyse it relative to each of these perspectives:

1. *The financial perspective.* Financial metrics could include cost:spend ratio.
2. *The business process perspective.* This perspective refers to internal business processes. Metrics based on this perspective allow the managers to know how well their business is running, and whether its products and services conform to customer requirements (the mission). Metrics could include: percentage of actions utilising electronic commerce, achievement of goals relating to sustainability, socioeconomic factors, etc.
3. *The customer perspective.* If customers are not satisfied, they will eventually find other suppliers that will meet their needs. Poor performance from this perspective is thus a leading indicator of future decline, even though the current financial picture may look good.
4. *The learning and growth perspective.* This perspective includes employee training and corporate cultural attitudes related to both individual and corporate self-improvement. In a knowledge–worker organisation, *people* – the only repository of knowledge – are the main resource. In the current climate of rapid technological change, it is becoming necessary for knowledge workers to be in a continuous learning mode. Government agencies often find themselves unable to hire new technical workers and at the same time is showing a decline in training of existing employees. This is a leading indicator of 'brain drain' that must be reversed. Metrics can be put into place to guide managers in focusing training funds where they can help the most. In any case, learning and growth constitute the essential foundation for success of any knowledge–worker organisation. Metrics could include percentage of employees satisfied with management, work environment, health and safety. Also the number of employees with recognised training and qualifications.

Generally, as with any management tool, the results obtained either through the adoption of KPIs or BSC techniques must be taken seriously and must be seen to be adopted. Understanding what a particular result really means is important in determining

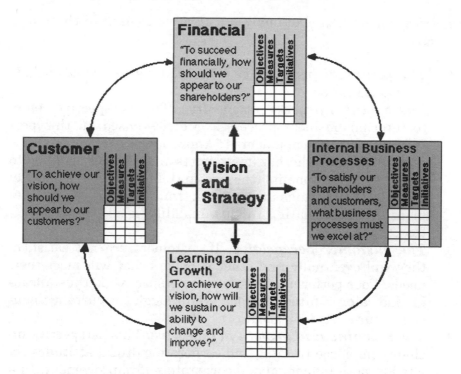

Figure 1.4 The balanced scorecard.

whether or not it is useful to the organisation. Data by itself is not useful information, but it can be when viewed from the context of organisational objectives, environmental conditions and other factors. Proper analysis is imperative in determining whether or not performance indicators are effective, and results are contributing to organisational objectives.

Conclusion

The construction industry still persists in the practice of rewarding bad behaviour. If a contract is delayed, all participants get their money apart from the client, who has to pay! There can be no doubt that the pressure for change within the UK construction industry and its professions, including quantity surveying, is unstoppable, and that the volume of initiatives in both the public and private sectors to try to engineer change grows daily. The last decade of the twentieth century saw a realignment of the UK's economic base. Traditional manufacturing industries declined while services industries prospered, but throughout this period the construction industry has remained relatively static, with a turnover compared to GDP of around 10 per cent. The construction industry is still, therefore, a substantial and influential sector and a major force in the UK economy. Perhaps more than any other construction profession, quantity surveying has repeatedly demonstrated the ability to reinvent itself and adapt to change.

Is there evidence that quantity surveyors are innovating and developing other fields of expertise? In 2004 a report was published by the RICS Foundation which came to the conclusions that there was evidence of innovation especially among the larger practices.

The report was based on a survey of 27 consultants from among the largest in the UK, ranked by the number of chartered quantity surveyors employed. The report concluded that there is a clear divide between the largest firms, each generating an income of more than £30 million per annum and the other firms surveyed. Several of the firms in the £5 – 15 million fee bands had recently made the transition from partnership to corporate status, while around half of the firms surveyed retained their traditional partnership structure. Of the private limited companies this had resulted in organisations of a very different shape with a flatter structure permitting more devolved responsibility and the potential for better communication throughout the organisation. The firms were asked to identify what percentage of fee income came from 'quantity surveying' services and all other fee-income generating services, the results indicated a significant diversification away from traditional quantity surveying services, as illustrated in Table 1.5.

The results indicate that in the case of the largest firms only just under 50 per cent of fee income came from quantity surveying

Table 1.5 Percentage of fee income from quantity suveying services

Annual fee income £ million	% from quantity surveying services		
	Mean	*Min*	*Max*
>30	49	29	80
20–30	5	5	5
15–20	–	–	–
10–15	63	25	90
5–10	66	36	86
<5	64	34	95

Source: RICS Foundation (2004).

services. The services being offered by the firms include: project management, legal services, taxation advice, value management and PFI consultancy.

The remainder of this book will attempt to review the new opportunities that are presenting themselves to the quantity surveyor in a swiftly changing global construction market. It is not the object of this book to proclaim the demise of the traditional quantity surveyor practice offering traditional quantity services – these will continue to be in demand – but rather to outline the opportunities that are now available for quantity surveyors to move into a new era offering a range of services and developing new expertise.

Bibliography

Agile Construction Initiative (1999). *Benchmarking the Government Client*. Stage Two Study. HMSO.

Auditor General (2000). *Modernising Construction*. HMSO.

Banwell, Sir H. (1964). *Report of the Committee on the Placing and Management of Contracts for Building and Civil Engineering Work*. HMSO.

Building Design Partnership (2004). *Learning from French Hospital Design*. Building Design Partnership.

Building/MTI (1999). *QS Strategies 1999, Volumes 1 and 2*. Building/Market Tracking International.

Burnside, K. and Westcott, A. (1999). *Market Trends and Developments in QS Services*. RICS Research Foundation.

Casl, B. and Winch, G.M. (1998). *Construction Benchmarking: An Internatinal Perspective*. Engineering and Physical Research Council.

Commission for Architecture and the Built Environment (2002). *Improving Standards of Design in the Procurement of Public Buildings*, London.

Construction Industry Computing Association (2000). *Architectural IT Usage and Training Requirements*. http://www.cica.org.uk

Cook, C. (1999). QS's in revolt. *Building*, 29 Oct, p. 24.

CRINE (1994). *Cost Reduction Initiative for a New Era*. United Kingdom Offshore Operators Association.

Department of the Environment, Transport and the Regions (1998). *Rethinking Construction*. HMSO.

Department of the Environment, Transport and the Regions (2000). *KPI Report For The Minister for Construction*. HMSO.

Edkins, A.J. and Winch, G.M. (1999). *Project Performance in Britain and France: the Case of Euroscan*, Barlett Research Paper 7.

Emmerson, Sir H. (1962). *Survey of Problems before the Construction Industries*. HMSO.

Financial News (2000). Report finds majors shunning traditional work. *Building*, 24 Nov, p. 21.

Goodall, J. (2000). *Is the British Construction Industry still suffering from the Napoleonic Wars?* Address to National Construction Creativity Club, London, 7 July 2000.

Hoxley, M. (1998). *Value for Money? The Impact of Competitive Fee Tendering on the Construction Professional Service Quality*. RICS Research.

Latham, Sir M. (1994). *Constructing the Team*. HMSO.

Leibfried, K.H.J. and Mc Nair, C. J. (1994). *Benchmarking: A Tool for Continuous Improvement*. Harper Collins.

Lema, N.M. and Price, A. D. F. (1995). Benchmarking performance improvement towards competitive advantage. *Journal of Management of Engineering* 11, 28–37.

Levene, Sir P. (1995). *Construction Procurement by Government*. An Efficiency Scrutiny, HMSO.

National Audit Office (2001). *Modernising Construction*, HMSO.

Pullen, L. (2001). What is best value in construction procurement? *Chartered Surveyor Monthly*, Feb.

Royal Institution of Chartered Surveyors (1971). *The Future Role of the Quantity Surveyor*. RICS.

Simon, Sir E. (1944). *The Placing and Management of Building Contracts*. HMSO.

Thompson, M.L. (1968). *Chartered Surveyors: The Growth of a Profession*. Routledge & Kegan Paul.

Watson, K. (2001). Building on shaky foundations. *Supply Management*, 23 Aug, pp. 23–20.

Winch, G.M. (1998). The Channel Fixed Link: Le project du Siecle. UMIST case study No. 450.

Websites

National Audit Office – http://www.nao.gov.uk
SIMAP – http://simap.eu.int
Tenders Electronic Daily (TED) – http://ted.eur-op.eu.int
Treasury – http://www.hm-treasury.gov.uk

2

Managing value.
Part 1: The supply chain

Introduction

This chapter examines the paths by which quantity surveyors can deliver a new range of added value services to clients, based upon increased client focus and a greater understanding of the function of built assets including, why new buildings are commissioned. As will be discussed, many clients who operate in highly competitive global markets, base their procurement strategies on the degree of added value that can be demonstrated by a particular strategy. In order to meet these criteria quantity surveyors must 'get inside the head' of their clients, fully appreciate their business objectives and find new ways to deliver value and thereby conversely remove waste from the procurement and construction process. The chapter will examine the application of manufacturing philosophies to the construction industry and the role of the quantity surveyor in the implementation of these approaches.

Does your client feel good?

The dictionary definition of procurement is 'the management of obtaining goods and services'. For many years quantity surveyors took this to mean appointing a contractor who submitted the lowest, that is to say cheapest tender price, based upon a bill of quantities and drawings, in competition with several other contractors. For public sector works, a low priced tender was almost guaranteed to win the contract, as public entities had to be seen to be spending public money prudently and would have had to develop a very strong case not to award a contract on the basis of the cheapest price. It is now clear that assembling an ad hoc list of six or so contractors, selected primarily on their availability, to tender for building work, for which they have very little detailed information, is not the best way to obtain value for money. In fact it does little more than reinforce the system in which contractors submit low initial prices secure in the knowledge that the contract will bring many opportunities to increase profit margins in the form of variation orders, claims for extensions of time/loss and expense or mismanagement by the design team. Pre-Latham, the over-riding ethos for procuring building works was to treat the supply chain with great suspicion, it was almost as if a cold war existed between client, design team and contractor/subcontractors. The emphasis in construction procurement has now swung away from the system described above towards systems that encourage partnerships and inclusion of the supply chain at an early stage – in fact to a point where the definition of procurement could be restated as 'obtaining value for money deals'. As discussed in Chapter 1, there can now be few individuals involved within the construction process who do not believe that the design, procurement and construction of new built assets has to become more efficient and client orientated. The evidence of wastage, in terms of materials, time and money not only in the short term, but also throughout the life cycle of a building, leaves the UK construction industry as well as its associated professions, including quantity surveyors, in an embarrassing position and open to criticism from all sides for participating in the production of such a low value product. Bernard Williams' study 'Building and Development Economics in the EC' demonstrated that the British construction industry was the least productive in Europe and as discussed in Chapter 1, this state of affairs appears to exist still. For whatever reason, the quantity surveyor and the traditional brick counting image enthusiastically fostered by so many within the industry, including the trade press, has also been the focus for this 'out of touch' image. For many years the quantity surveyor has been seen as the accountant to the construction industry, a knight in shining armour, safeguarding the client to ensure that they receive a building as close as possible to the initial agreed target price, although in practice, this has been seldom achieved. Traditionally, a target cost has been set by the quantity surveyor, in discussion with the client at the outset and then the process has been worked backwards squeezing in turn the contractors, subcontractors and suppliers in order to keep within this target cost. The squeezing increases in direct proportion to how far down the supply chain the organisation comes (see Figure 2.1).

The consequences of the supply chain squeeze illustrated in Figure 2.1 are low profit margins, lack of certainty and continuity for suppliers, delays in production, lack of consideration of through life costs, suboptimal functionality and rampant waste. In turn low profits ensure that few, if any, resources can be channelled into research, technological improvement, or quality assurance procedures. Construction productivity lags behind that of manufacturing and yet manufacturing has been a reference point and a source of innovation in construction for many decades, for example, industrialised building, currently undergoing a resurgence in interest in the UK, and the use of computer-aided design come directly from the manufacturing sector. However, while some innovations have crossed the divide from manufacturing to construction, there has

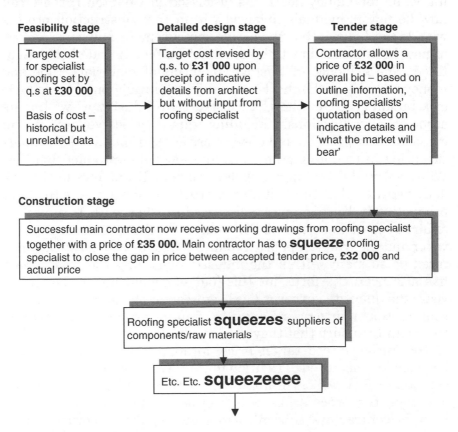

Figure 2.1 Supply chain squeeze.

been little enthusiasm for other production philosophies. However, new manufacturing industry based approaches to supply chain management are now being introduced into the construction industry that enable the early involvement of suppliers and subcontractors in a project with devolved responsibility for design and production of a specific section of a building, with predicted and guaranteed through life costs, for periods of up to 35 years. The benefits of this new approach for clients include the delivery of increased functionality at reduced cost and, for the supply chain members, certainty, less waste and increased profits.

Supply chain relationships and management

A construction project organisation is usually a temporary organisation designed and assembled for the purpose of the particular project. It is made up of different companies and practices, which have not necessarily worked together before and which are tied to the project by means of varying contractual arrangements. This is what has been termed a temporary multi-organisation. Its temporary nature extends to the workforce, which may be employed for a particular project, rather than permanently. These traditional design team/supply chain models are the result of managerial policy aimed at sequential execution and letting out the various parts of the work at apparently lowest costs. The problems for process control and improvement that this temporary multi-organisation approach produces are related to:

- Communicating data, knowledge and design solutions across the organisation
- Stimulating and accumulating improvement in processes that cross the organisational borders
- Achieving goal congruity across the project organisation
- Stimulating and accumulating improvement inside an organisation with a transient workforce.

The following quotation, attributed to Sir Denys Hinton speaking at the RIBA in 1976 seems to sum up the traditional attitude of building design teams:

> ... the so-called building team. As teams go, it really is rather peculiar, not at all like a cricket eleven, more that a scratch bunch

consisting of one batsman, one goal keeper, a pole vaulter and a polo player. Normally brought together for a single enterprise, each member has different objectives, training and techniques and different rules. The relationship is unstable, and with very little functional cohesion and no loyalty to a common end beyond that of coming through unscathed.

Most of what is encompassed by the term 'supply chain management' was formerly referred to by other terms such as 'operations management', but the coining of a new term is more than just new management speak, it reflects the significant changes that have taken place across this sphere of activity. These changes result from changes in the business environment. Most manufacturing companies are only too aware of such changes, increasing globalisation, savage price competition, increased customer demand for enhanced quality and reliability, etc. Supply chain management was introduced in order that manufacturing companies could increase their competitiveness in an increasingly global environment as well as their market share and profits by:

- Minimising the costs of production on a continuing basis
- Introducing new technologies
- Improving quality
- Concentrating on what they do best.

The contrast between traditional approaches and supply chain management is summarised in Figure 2.2.

The quantity surveyor as supply chain manager

What is the driving force for the introduction of supply chain management into the UK construction industry? As described in Chapter 1, the CRINE initiative in the oil industry was the result of the collapse of world oil prices to $13 a barrel in 1992; however, in construction very little impetus has come from the industry, it is clients that are the driving force. As discussed in Chapter 6, unlike other market sectors, because the majority of organisations working in construction are small, the industry has no single organisation to champion change. When Latham called for a 30 per cent reduction in costs, the knee jerk response from some quarters of the profession and industry was that cost = prices and therefore it

Supply chain management	Traditional model
Target cost	Competitive tender
Cost transparency	Fixed price
Integrated teams	Fragmentation
Shared benefits for improved delivery	Penalties for non-delivery

Figure 2.2 Supply chain and traditional management approaches compared.

was impossible to reduce the prices entered in the bill of quantities by this amount, therefore the target was unrealistic and unachievable. However, this was not what Latham was calling for, as will be demonstrated in the following paragraphs. Reducing costs goes far beyond cutting the prices entered in the bill of quantities, if it ever did, it extends to the re-organisation of the whole construction supply chain in order to eliminate waste and add value. The immediate implications of supply chain management are:

- Key suppliers are chosen on criteria, rather than job by job on competitive quotes
- Key suppliers are appointed on a long-term basis and proactively managed
- All suppliers are expected to make sufficient profits to reinvest.

How many quantity surveyors have asked themselves this question at the outset of a new project;

What does value mean for my client?

In other words, in the case of a new plant to manufacture say, pharmaceutical products, what is the form of the built asset that will deliver value for money, over the life cycle of the building for that particular client? For many years, whenever clients have voiced their concerns about the deficiencies in the finished product, all too often the patronising response from the profession has been to accuse the complainants of a lack of understanding in either design or the construction process or both. The answer to the value question posed above will, of course, vary between clients, a large multinational manufacturing organisation will have a different view of value to a wealthy individual commissioning a new house, but it helps to illustrate the revolution in thinking and attitudes that must take place. In general, the definition of value for a client is 'design to meet a functional requirement for a through-life cost'. Quantity surveyors are increasingly developing better client focus, because only by knowing the ways in which a particular client perceives or even measures value, whether in a new factory or a new house, can the construction process ever hope to provide a product or service that matches these perceptions. Once these value criteria are acknowledged and understood quantity surveyors have a number of techniques, described in this chapter, at their disposal in order to deliver to their clients a high degree of the feel good factor.

Not all of the techniques are new, many practising quantity surveyors would agree that the strength of the profession is expertise in measurement and in supply chain management there is a lot to measure, for example:

- Measure productivity – for benchmarking purposes
- Measure value – demonstrating added value
- Measure out-turn performance – not the starting point
- Measure supply chain development – are suppliers improving as expected?
- Measure ultimate customer satisfaction – customers at supermarket, passengers at airport terminal, etc.

Of course, measuring value is extremely difficult. CIRIA (Construction Industry Research and Information Association) and Loughborough University have been working along with clients' group, the Construction Round Table, to develop a client's guide for standardisation and pre-assembly. One of the tenets of this work is the need to measure success and in particular to move this measurement beyond the obvious cash-related benefits. The team has developed specific 'measures' for both standardisation and pre-assembly similar to the Key Performance Indicators (KPIs) that are in use for other aspects in construction. Such generic measures can be useful for the industry-wide comparison but have limited use in measuring project-specific issues. Therefore, in addition to these the team produced methods of measuring benefits related to the drivers and constraints of the particular project. A related project called IMMPREST (Interactive Model for Measuring the benefits from Standardisation and Pre-assembly) is developing these tools further by comparing different methods of producing buildings or building elements against the following criteria:

- Monetary benefits
- Programme benefits
- Quality benefits
- Logistical and operational benefits
- Health and safety benefits
- Environmental and sustainability benefits
- Organisational benefits
- Other benefits.

What is a supply chain?

Before establishing a supply chain or supply chain network, it is crucial to understand fully the concepts behind, and the possible components of, a complete and integrated supply chain. The term 'supply chain' has become used to describe the sequence of processes and activities involved in the complete manufacturing and distribution cycle – this could include everything from product design through materials and component ordering through manufacturing and assembly until the finished product is in the hands of the final owner. Of course, the nature of the supply chain varies from industry to industry. Members of the supply chain can be referred to as upstream and downstream supply chain members (Figure 2.3).

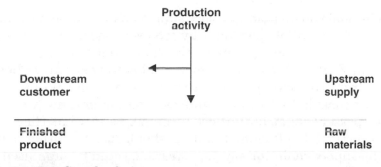

Figure 2.3 Supply chain.

Supply chain management, which has been practised widely for many years in the manufacturing sector, therefore refers to how any particular manufacturer involved in a supply chain, manages its relationship both up and downstream with suppliers to deliver cheaper, faster and better. In addition good management means creating a safe commercial environment, in order that suppliers can share pricing and cost data with other supply team members.

The more efficient or lean the supply chain, the more value is added to the finished product. As if to emphasise the value point, some managers substitute the word 'value' for 'supply' to create the value chain. In a construction context supply chain management involves looking beyond the building itself and into the process, components and materials which make up the building. Supply chain management can bring benefits to all involved, when applied to the total process which starts with a detailed definition of the client's business needs which can be provided through the use of value management and ends with the delivery of a building which provides the environment in which those business needs can be carried out with maximum efficiency and minimum maintenance and operating costs. In the traditional methods of procurement, the supply chain does not understand the underlying costs, hence suppliers are selected by cost and then squeezed to reduce price and whittle away profit margins. An approach characterised by:

• Bids based on designs to which suppliers have no input
• No buildability
• Low bids always won
• Unsustainable – costs recovered by other means
• Margins low, so no money to invest in development
• Suppliers distant from final customer so took limited interest in quality.

The traditional construction project supply chain can be described as a series of sequential operations by groups of people who have no concern about the other groups or the client and/or end user. For example, in Figure 2.4 the end user is deliberately shown as being detached from the supply chain activity as typically no one has thought to consult this group, who may be patients in a health centre or passengers at an airport terminal, as to what they perceive a functionally efficient building should be. In rather more detail, the supply chain for say, fenestration, could be organised as in Figure 2.5.

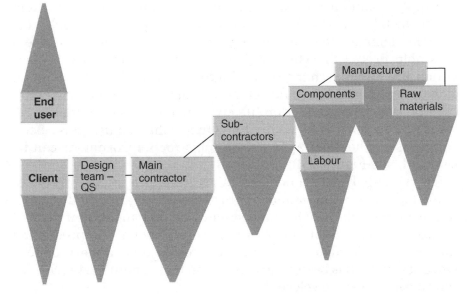

Figure 2.4 Traditional construction supply chain.

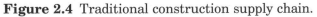

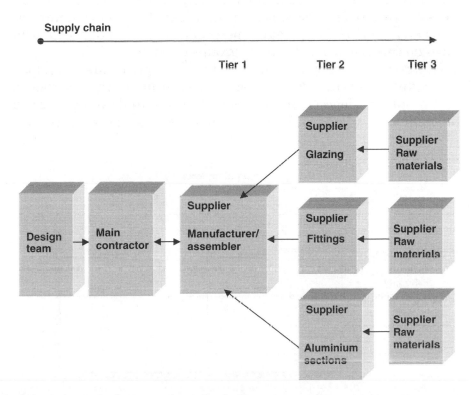

Figure 2.5 Fenestration supply chain.

Supply chains are unique, but it is possible to classify them generally by their stability or uncertainty on both the supply side and the demand side. On the supply side, low uncertainty refers to stable processes, while high uncertainty refers to processes which are rapidly changing or highly volatile. On the demand side, low uncertainty would relate to functional products in a mature phase of the production life cycle, while high uncertainty relates to innovative products. Once the chain has been catagorised, the most appropriate tools for improvement can be selected (see Figure 2.6).

On the supply side, low uncertainty refers to stable processes, while high uncertainty refers to processes which are rapidly changing or highly volatile. On the demand side, low uncertainty would relate to functional products in a mature phase of the production life cycle, while high uncertainty relates to innovative products. Once the chain has been catagorised, the most appropriate tools for improvement can be selected.

The traditional supply chain arrangement in Figure 2.4 is characterised by lack of management, little understanding between tiers of other tiers, functions or processes and lack of communications and a series of sequential operations by groups of people who have no concern about the other groups or client.

However, with supply chain management, prices are developed and agreed, subject to an agreed maximum price with overheads and profit ring fenced. All parties collaborate to drive down costs and enhance value with, for example the use of an incentive

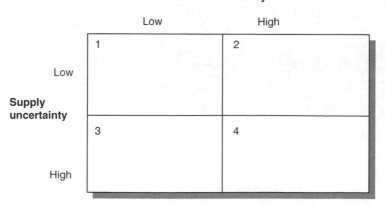

Figure 2.6 Supply and demand uncertainty.

scheme.With cost determined and profit ring fenced, waste can now be attacked to bring down price and add value and there is an emphasis on continued improvement.

- As suppliers account for 70–80 per cent of building costs, they should be selected on their capability to deliver excellent work at competitive cost
- Suppliers should be able to contribute new ideas, products and processes
- Build alliances outside the project
- Waste and inefficiency can be continuously identified and driven out.

Figure 2.7 illustrates the seven principles of supply chain management as suggested by the *Building Down Barriers Handbook*.

The philosophy of integrated supply chain management is based upon defining and delivering client value through established supplier links that are constantly reviewing their operation in order to

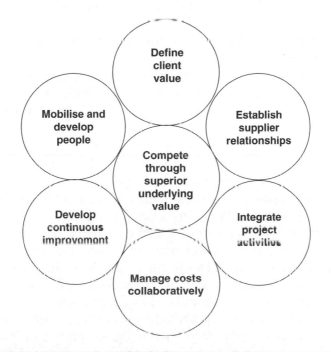

Figure 2.7 Principles of supply chain management
(Source: Defence Estates/Building Research Establishment).

improve efficiency. There are now growing pressures to introduce these production philosophies into construction and it is quantity surveyors with their traditional skills of cost advise and project management who can be at the forefront of this new approach. For example, the philosophy of Lean Thinking, which is based on the concept of the elimination of waste from the production cycle, is of particular interest in the drive to deliver better value. In order to utilize lean thinking philosophy, the first hurdle that must be crossed is the idea that construction is a manufacturing industry which can only operate efficiently by means of a managed and integrated supply chain. At present the majority of clients are required to procure the design of a new building separately from the construction. However, as the subsequent delivery often involves a process where sometimes as much as 90 per cent of the total cost of the completed building is delivered by the supply chain members there would appear to be close comparisons with, say, the production of a motor car or an airplane.

The basics of supply chain management can be described as:

1. Determine which are the strategic suppliers and concentrate on these key players as the partners who will maximise added value
2. Work with these key players to improve their contribution to added value
3. Designate these key suppliers as the 'first tier' on the supply chain and delegate to them the responsibility for the management of their own suppliers, the 'second tier' and beyond.

To give this a construction context, the responsibility for the design and execution of, say, mechanical installations could be given to a 'first tier' engineering specialist. This specialist would in turn work with its 'second tier' suppliers, as well as with the design team to produce the finished installation. Timing is crucial as first tier partners must be able to proceed confident that all other matters regarding the interface of the mechanical and engineering installation with the rest of the project have been resolved and that this element can proceed independently. Although at least one food retail organisation using supply chain management for the construction of its stores, still places the emphasis on the tier partners to keep themselves up to date with progress on the other tiers, as any other approach would be incompatible with rapid timescales that are demanded.

Despite the fact that on the face of it certain aspects of the construction process appear to be prime candidates for this approach, the biggest obstacles to be overcome by the construction industry in adopting manufacturing industry style supply chain management are:

1. Unlike in manufacturing, the planning, design and procurement of a building is at present separated from its construction or production
2. The insistence that unlike an airplane or motorcar, every building is bespoke, a prototype, and therefore is unsuited to this type of model or for that matter any other generic production sector management technique. This factor manifests itself by:

    *   Geographical separation of sites that causes breaks in the flow of production
    *   Discontinuous demand
    *   Working in the open air, exposed to the elements. Can there be any other manufacturing process, apart from shipbuilding that does this?

3. Reluctance by the design team to accept early input from suppliers and subcontractors and unease with the blurring of traditional roles and responsibilities.

There is little doubt that the first and third hurdles are the result of the historical baggage outlined in Chapter 1 and that, given time, they can be overcome. However, the second hurdle does seem to have some validity, despite statements from the proponents of production techniques, buildings are not unique and that commonality even between apparently differing building types is as high as 70 per cent (Ministry of Defence, *Building Down Barriers*). Interestingly though, one of the main elements of supply chain management, Just in Time (JIT), was reported to have started in the Japanese shipbuilding industry in the mid-1960s, the very industry that opponents of JIT in construction quote as an example where, like construction, supply chain management techniques are inappropriate. Therefore, the point at which any discussion of the suitability of the application of supply chain management techniques to building has to start with the acceptance that construction is a manufacturing process, which can only operate efficiently by means of a managed and integrated supply chain. One fact is undeniable – at present the majority of clients are required to procure the design of a new building separately from the construction. Until comparatively recently, international competition, which in manufacturing is a major influencing factor, was relatively sparse in domestic construction of major industrialised countries.

Adding value and minimising waste

One of the best researched industries is car manufacturing. Lean car production is characterised as using less of everything compared with mass production; half of the human effort, half of the manufacturing space, half of the engineering hours to develop a new product in half the time. The competitive benefits created by means of the new approach seem to be remarkably sustainable, however, with the exception of quality methodologies, this new philosophy is little known in construction. The ideas of the new production philosophy first originated in Japan in the 1950s. The most prominent application, as pioneered by Taiichi Ohno (1912–1990), was the Toyota production system and central to the system was the single-minded determination to eliminate waste through, amongst other techniques, co-operation with suppliers. Ohno was credited with the identification of waste, defined as activities that create no value, as follows: *defects* in products, *overproduction* of goods not needed, *inventories* of goods awaiting further processing or consumption, unnecessary *processing,* unnecessary *movement* of people, unnecessary *transport* of goods and *waiting* by downstream activities for processes to finish on an upstream activity. In 1996, Womack and Jones added an eighth: the design of goods and services *unsuitable* for users' needs. How many of the items highlighted above can be identified in a typical construction project – Four? Five? More?

Simultaneously, under the guidance of American consultants, quality issues became another focus for Japanese industry. By the beginning of the 1990s the new production philosophy, which is known by several different names including world-class manufacturing, the new production system and lean production was in widespread use. In 1998, the term 'lean construction' was introduced by DETR and central to this new direction are the concepts of Just in Time (JIT) and total quality control (TQC). To the original lean production philosophy, many other related concepts have been added, too numerous to describe here; however, three have been included, as particularly relevant for the quantity surveyor in the drive for added value in construction: namely value engineering/management, concurrent engineering and continuous improvement. The core of the new production philosophy is in the observation that there are two kinds of phenomena in all new production systems:

- Conversions
- Flows.

Conversions versus flows

The conventional approach to a construction project based on conversions is illustrated in Figure 2.8.

The conversion approach to construction revolves around the principle that buildings are conceived as sets of operations, which are controlled operation by operation, for least cost. However straightforward the conversion appears, it masks particularly in the construction process, inherent waste due to:

1. Rework due to design or construction errors
2. Non-value-adding services in the material and work flows, such as waiting and handling and double handling
3. Inspecting, duplicating activities and accidents.

The primary focus therefore in the design of new built assets is on minimising value loss, whereas in construction it is on minimising waste.

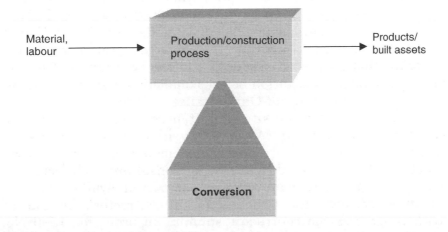

Figure 2.8 Conversion process.

The following statistics are taken from various reports from CIRIA and Movement for Innovation:

- Every year in the UK, approximately 13 million tonnes of construction materials are delivered to site and thrown away, unused
- Ten per cent of products are wasted through over supply, costing £2.4 billion per annum
- Another £2.4 billion per annum is wasted in stockpiling materials
- £5 billion per annum is squandered by the misuse of materials.

The lean construction tool box

The following approaches are available to achieve lean construction:

1. Just-in-time production
2. Total quality control/continuous improvement
3. Value management
4. Concurrent engineering.

1. Just-in-time production

This is the starting point for Ohno and the Toyota revolution. This so-called pull type production process is based on the principle that good communications with suppliers ensures than production is initiated by actual demand, rather than by plans based on forecasts and contrasts to the push technology where large volumes of materials and components are produced, transported and stored ahead of demand and requirements. The driving idea is the reduction or elimination of work in progress, that is large stocks of goods and materials that have been produced and are awaiting further processing and/or transportation before completion, thereby eliminating two or possibly three of Ohno's causes of waste.

For example, delays on site can often be the result of the unavailability of materials. A quantity surveyor has prepared a schedule of the door and window requirements for a new hospital, which has been designed by a firm of consultant architects, but without reference to material and component availability. The schedule becomes part of the successful contractor's bid and as work progresses the contractor submits an order for 1500 No. 2400 × 1200 mm double-glazed window units. The different response paths to this order are illustrated in Figure 2.9, the first path reflecting the traditional approach to production, the so-called push method where the supplier is not able to fill the order because of lack of units and inflexible pre-set production schedules, the second path reflecting the pull method, that is able to react more quickly to demand. In the centre of the diagram between the two methods of production are some of the elements of

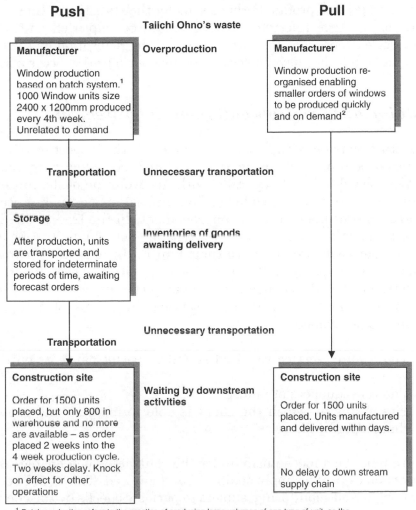

Push

Pull

Taiichi Ohno's waste

Manufacturer

Window production based on batch system.[1] 1000 Window units size 2400 x 1200mm produced every 4th week. Unrelated to demand

Overproduction

Manufacturer

Window production re-organised enabling smaller orders of windows to be produced quickly and on demand[2]

Transportation

Unnecessary transportation

Storage

After production, units are transported and stored for indeterminate periods of time, awaiting forecast orders

Inventories of goods awaiting delivery

Unnecessary transportation

Transportation

Construction site

Order for 1500 units placed, but only 800 in warehouse and no more are available – as order placed 2 weeks into the 4 week production cycle. Two weeks delay. Knock on effect for other operations

Waiting by downstream activities

Construction site

Order for 1500 units placed. Units manufactured and delivered within days.

No delay to down stream supply chain

[1] Batch production refers to the practice of producing large volumes of one type of unit, as the cheapest method of production. Volumes based on forecasts not demand.
[2] Just-in-time production pioneered by Toyota. Production becomes more responsive and focused, based on small lot production and reduced set up times.

Figure 2.9 Lean building component production.

waste identified by Taiichi Ohno, each time waste is removed from the process, value is added! This example illustrates the potential problems with just one component of a building project and how traditionally many of the components commonly used in construction have large elements of waste built into their production. However, this sort of scenario can be encountered many times during the construction of a project.

Every time this waste is removed from the supply chain, value is added to the process leading to lower costs, shorter construction periods and greater profits. Perhaps some of the statistics of time and material wastage referred to earlier could be improved with the wide-scale adoption of JIT techniques and the realisation by the design team that they have to design and manage a project accordingly.

Total quality control/continuous improvement

The starting point of the quality movement is the inspection of raw materials and products using statistical methods. Quality methodologies have developed in parallel with the evolution of the concept of quality. The focus has changed from inspection, through process control to continuous process improvement. In the traditional approach to quality, no special effort is made to eliminate defects, errors or omissions or to reduce their impact. In numerous studies from different countries the cost of poor quality has been found to be between 10 and 20 per cent of total project costs. Agreements for the procurement of construction services by means of a supply chain should include:

- The level of service required stated in performance or output terms
- The consequences of failure
- The means by which the client is able to measure the supply chain's performance.

There must be a mechanism under the contract which enables the performance of the supply chain to be monitored. This could be in the form of self-monitoring, subject to periodic checks by the client; open book accounting is normally required for this to work effectively. In Chapter 1 the process of benchmarking was described, but what and how to benchmark the supply chain? One approach is to introduce risk reward schemes that enable the supply chain to share in the benefit of the works being completed under the target cost. Conversely, the supply chain will share the client's downside if the final cost exceeds the target. This approach differs from the establishment of a guaranteed maximum price and this is thought by some practitioners to be at odds with the collaborative team approach.

Figure 2.10 illustrates a scheme related to supply chain service providers sharing the risk and reward based on the target cost, in which the dead band has been agreed beforehand at £1 million. This band is not fixed and will reflect the confidence of the parties in the target value of a particular project. Similar arrangements can be made for completion targets as illustrated in Figure 2.11.

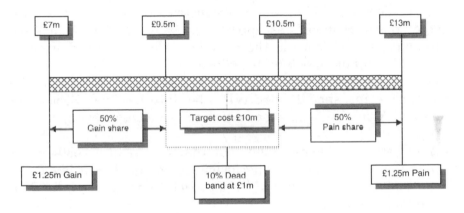

Figure 2.10 Target cost scheme.

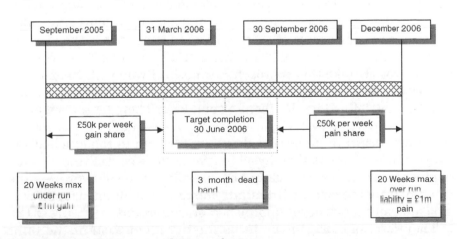

Figure 2.11 Target completion scheme.

Value analysis, engineering, management

Central to the goal of delivering built assets which meet the functional and operational needs of a client are the techniques of value engineering and value management. Developed first in the USA for the manufacturing and production sectors, by Lawrence D. Miles, in the immediate post Second World War era as value analysis, later rebadged as value engineering/management, this approach is now widely practised by UK quantity surveyors in both public and private sectors. To quote Robert N. Harvey, one time manager of capital programmes and value management for the Port Authority of New York and New Jersey; *'Value Engineering is like love – until you've experienced it you just can't begin to understand it'*. In the early 1990s, the Port Authority conducted value engineering workshops on nearly $1 billion worth of construction projects. The total cost of the workshops was approximately $1 million, a massive statement of confidence in the technique that paid off delivering nearly $55 million in potential savings.

For a somewhat more objective view of the process, perhaps the reference point should be SAVE, the International Society of American Value Engineers' definition of value engineering is:

> A powerful problem-solving tool that can reduce costs while maintaining or improving performance and quality. It is a function-oriented, systematic team approach to providing value in a product or service.

The philosophy of value engineering/management is a step change from the traditional quantity surveying belief that delivering value is based on the principle of cutting costs to keep within the original budget – what was and still is euphemistically referred to as, cost reconciliation. Unlike this approach, the basis of value management is to analyse, at the outset, the function of a building, or even part of a building, as defined by the client or end user. Then, by the adoption of a structured and systematic approach, to seek alternatives and remove or substitute items that do not contribute to the efficient delivery of this function, thereby adding value. The golden rule of value engineering/management is that as a result of the value process the function(s) of the object of the study should be maintained and if possible enhanced, but never diminished or compromised (Figure 2.12).

Therefore, once again the focus for the production of the built asset is a client's perception of value. Perhaps before continuing

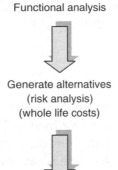

Functional analysis

Generate alternatives
(risk analysis)
(whole life costs)

Develop alternative
solutions

Figure 2.12 Value process.

much further the terms associated with various value methodologies should be explained. The terms in common usage are value analysis, value engineering and value management.

Value analysis

The name adopted by Lawrence D. Miles for his early studies and defined as an organised approach to the identification and elimination of unnecessary cost.

Value engineering

The name adopted in 1959 by SAVE when it was established, to formalise the Miles' approach. The term is widely used in North America and the essential philosophy of VE is 'a disciplined procedure directed towards the achievement of necessary function for minimum cost without detriment to quality, reliability, performance or delivery'. As if to emphasise the importance now being placed on value engineering in 2000 Property Advisors to the Civil Estate (PACE) introduced an amendment to GC/Works/1 – Value Engineering Clause 40(4). The amendment states;

The Contractor shall carry out value engineering appraisals throughout the design and the construction of the works to identify the function of the relevant building components and to provide the necessary function reliability at the lowest possible costs. If the Contractor considers that a change in the

Employer's Requirements could effect savings, the Contractor shall produce a value engineering report.

Value management

Value management involves considerably more emphasis on problem-solving as well as exploring in depth functional analysis and the relationship between function and cost. It also incorporates a broader appreciation of the connection between a client's corporate strategy and the strategic management of the project. In essence, value management is concerned with the 'what' rather than the 'how' and would seem to represent the more holistic approach now being demanded by some UK construction industry clients, i.e. to manage value. The function of value management is to reduce total through life costs comprising initial construction, annual operating, maintenance and energy costs and periodic replacement costs, without affecting and while indeed improving performance and reliability and other required design parameters. It is a function oriented study and is accomplished by evaluating functions of the project and its subsystems and components to determine alternative means of accomplishing these functions at lower cost. Using value management, improved value may be derived in three predominant manners:

1.   Providing for all required functions, but at a lower cost
2.   Providing enhanced functions at the same cost
3.   Providing improved function at a lower cost – the Holy Grail.

Amongst other techniques, value management uses value engineering study or workshop, that brings together a multidisciplinary team of people, independent to the design team, but who own the problem under scrutiny and have the expertise to identify and solve it. A value engineering study team works under the direction of a facilitator, who follows an established set of procedures (for example, the SAVE Value Methodology Standard, as in Figure 2.13) to review the project, making sure the team understands the client's requirements and develops a cost-effective solution. Perhaps the key player in a VE study is the facilitator or value management practitioner, who must within a comparatively short time ensure that a group of people works effectively together. People like Alphonse Dell'Isola, the Washington DC based practitioner, who

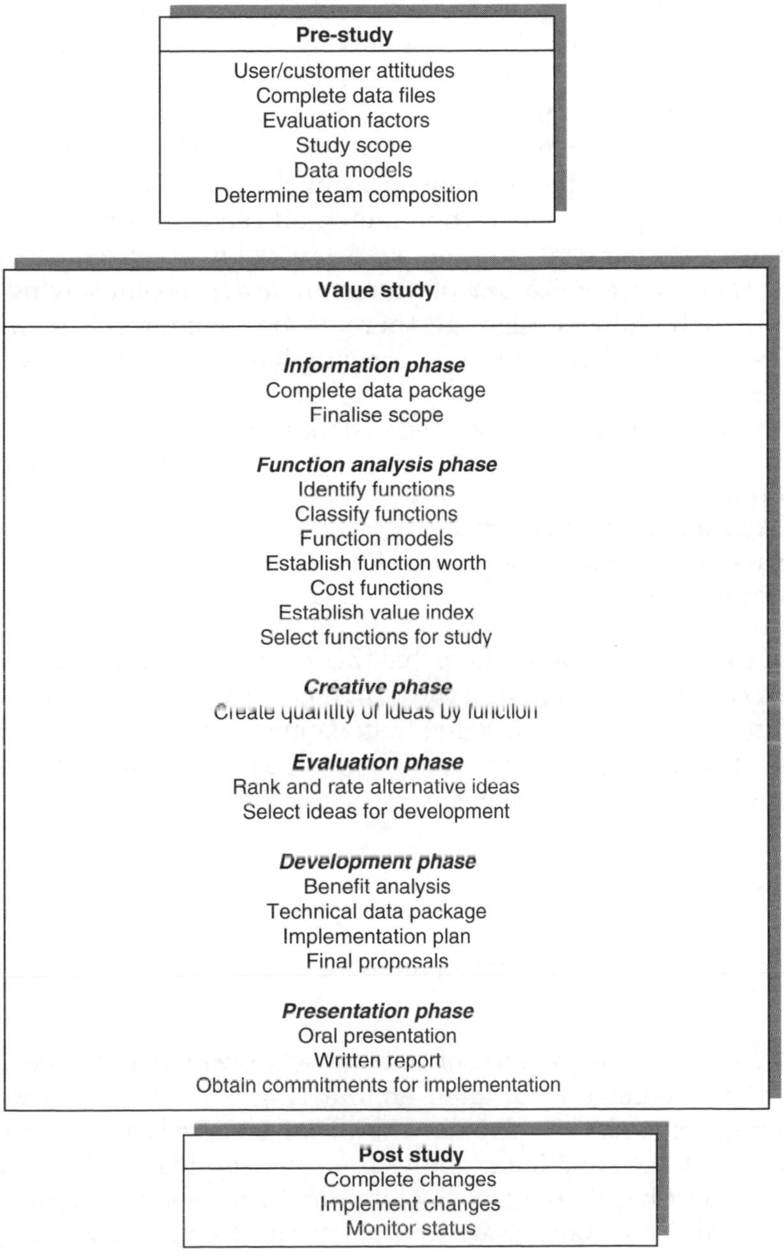

Pre-study
User/customer attitudes
Complete data files
Evaluation factors
Study scope
Data models
Determine team composition

Value study

Information phase
Complete data package
Finalise scope

Function analysis phase
Identify functions
Classify functions
Function models
Establish function worth
Cost functions
Establish value index
Select functions for study

Creative phase
Create quantity of ideas by function

Evaluation phase
Rank and rate alternative ideas
Select ideas for development

Development phase
Benefit analysis
Technical data package
Implementation plan
Final proposals

Presentation phase
Oral presentation
Written report
Obtain commitments for implementation

Post study
Complete changes
Implement changes
Monitor status

Figure 2.13 Value management methodology (Source: SAVE International Society of American Value Engineers).

rose to be an icon in value management circles and have helped
SAVE to prove their claim that value management produces
savings of 30 per cent of the estimated cost for constructing a project and that for every pound invested in a VE study, including participants' time and implementation costs, £10 is saved. Certainly,
organisations which have introduced VE into their existing procurement process, for example previously publicly owned water
companies, London Underground, etc., all report initial savings of
around 10–20 per cent. In some respects, value management is no
more than the application of the standard problem-solving approach to building design. If there is one characteristic which
makes VM/VE distinctive it is the emphasis given to functional
analysis.

The techniques that can be used to define and analyse function are:

- Value trees
- Decision analysis matrix
- Functional analysis system technique (FAST) diagrams
- Criteria scoring.

Once the function of an item has been defined then the cost or
worth can be calculated and the worth/cost ratio scrutinised to determine value for money. Value management can therefore be said
to be a holistic approach to managing value that includes the use of
value engineering techniques.

The process

The theory of value management is – buy function, don't buy
product.

Whilst it is not the purpose of this book to give a detailed description of every stage of a value engineering workshop it is worth
spending some time to explain the process as well as examine in
more detail the functional analysis phase, see Figure 2.13. Value
management has its roots in the manufacturing sector where it has
been around for many years and there can be problems applying
the approach to the construction of a new building. Nevertheless,
valuable insights into the functions behind the need for a new
building and what is needed to fulfil these functions can flow from
a value engineering workshop, even if it is not as lengthy and detailed as the traditional 40 hours programme used in North

America. If nothing else, it may be the only time in the planning and construction of a project when all the parties – client, end user, architect and quantity surveyor – sit down together to discuss the project in detail.

For example, let's consider a hypothetical project where a value engineering workshop is used, the construction of a new clinic for the treatment of cancer. There are numerous variations and adaptations of, not only the approach to conducting a value engineering workshop, but also the preparation of items like the FAST diagram (see Figure 2.14). To illustrate the procedure, the following example

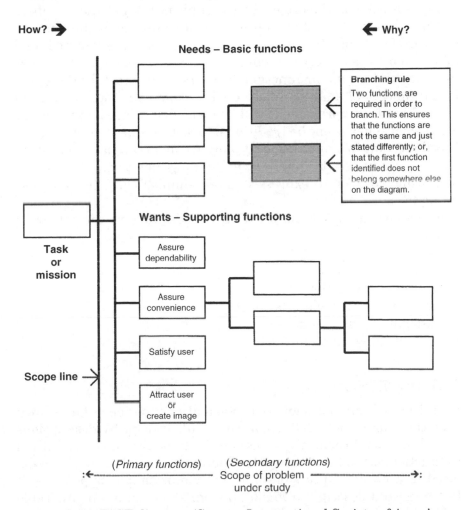

Figure 2.14 FAST diagram (Source: International Society of American Value Engineers).

is based on a classic, 40-hour, 5-day, value engineering workshop, as this presents a more proven and pragmatic approach than some of the latter day variations of value engineering. The workshop team is made up of six to eight experts from various design and construction disciplines, who are not affiliated to the project, as it has been found that the process is not so vigorous if in-house personnel are used. In addition, an independent facilitator is recommended as they have also been proven to be less liable to compromise on the delivery of any recommendations. The assembled team then commences the workshop, following the steps of the SAVE methodology (see Figure 2.13). At the start of the week, the group is briefed on the project by the clinic personnel and members of the design and construction team and the scope of the study is defined. Costs of the project are also carefully examined and analysed using a variety of techniques, as well as compared to other facilities with a similar function. The first major task in the study is the functional analysis phase during which the most beneficial areas for value improvement will be identified. Whilst unnecessary cost removal has been the traditional target for quantity surveyors, it is important to emphasise that more frequently today value studies are conducted to improve a building performance without increasing cost, or to express it more simplistically, to maximise 'bang for buck'.

Functional analysis using a FAST model follows the following steps:

* Defining function
* Classifying function
* Developing function relationships
* Assigning cost to function
* Establishing function worth.

Defining function

Definition of function can be problematic; experience has shown that the search for a definition can result in lengthy descriptions that do not lend themselves to analysis. In addition, the definition of function has to be measurable. Therefore, a method has been devised to keep the expression of a function as simple as possible; it is a two word description made up from a verb and a noun. Table 2.1 lists typical verbs and nouns; more comprehensive lists are readily available.

Table 2.1 Typical verbs and nouns used in functional analysis

Verbs		Nouns	
Amplify	Limit	Area	Power
Attract	Locate	Corrosion	Protection
Change	Modulate	Current	Radiation
Collect	Move	Damage	Repair
Conduct	Protect	Density	Stability
Contain	Remove	Energy	Surface
Control	Rotate	Flow	Vibration
Enclose	Secure	Fluid	Voltage
Filter	Shield	Heat	Volume
Hold	Support	Insulation	Weight

On first sight this approach may appear to be contrived, but it has proved effective in pinpointing functions. It is not cluttered with superfluous information and promotes full understanding by all members of the team regardless of their knowledge or technical backgrounds. For example, identify treatment, assess condition, diagnose illness (Figure 2.15).

Classifying function

In order to establish some sort of hierarchy, functions are classified into primary or basic function and supporting functions. Basic functions or needs are functions that make the project or service work, if omitted it would impact on the effectiveness of the completed project. Out of the list of basic functions emerges the highest order function, that can be defined as the overall reason for the project and meets the overall needs of the client. This function is placed to the left of the scope line on the FAST diagram (Figure 2.15). The second grouping of functions, supporting functions, may in a majority of cases contribute nothing to the value of a building. Supporting functions generally fall into the following categories:

- Assure dependability
- Assure convenience
- Satisfy user
- Create image.

At first glance these categories may seem to have little relevance to construction-related activities, until it is understood that for example, the 'create image' heading includes items such as aesthetic aspects, overall appearance, decoration and implied performance, such as reliability, safety, etc. Items that in themselves are not vital for the integrity of the project, but nevertheless may be high on the client's list of priorities.

Developing functional relationships

Functional analysis system technique (FAST) models are a method of depicting functional relationships (see Figures 2.14 and 2.15). The model works both vertically and horizontally by first determining the highest order function, called the task or mission, that is positioned to the left of the vertical scope line. By working from the left and asking the question 'HOW?' and employing the verb/noun combination and working from the right asking the question 'WHY?', the functions and their interrelationship can be mapped and their value allocated at a later phase.

Assigning cost to function

Conventionally, project costs are given in a detailed cost plan, where the actual costs of labour materials and plant are calculated and shown against an element, as in the Building Cost Information Service's standard list. (Table 2.2)

Value engineering is based on the concept that clients buy functions, not materials or building components, as defined and expressed by their user requirements. Therefore, splitting costs among the identified functions, such as a FAST diagram shows how resources are spent in order to fulfil these functions. Costs can then be viewed from the perspective of how efficiently they deliver the function. Obviously, the cost of each element can cover several functions – for example, the element BCIS Ref 2G Internal Walls and Partitions may contribute to the delivery of several functions of the project. It is therefore necessary at the outset to study the

Table 2.2 Elemental costs (Building Cost Information Service)

Element	Total cost of element (£)	Cost per m² of gross floor area (£)	Element unit quantity	Element unit rate (£)
2G Internal walls and partitions	430283.00	45.00	8025 m²	53.62

cost plan and to allocate the costs to the appropriate function (see Table 2.3). A similar exercise is carried out until all of the project costs are allocated to functions.

Establishing function worth

The next step is to identify which of the functions contains a value mismatch, or in other words seems to have a high contribution to the total project cost in relation to the function that it performs. Following on from this the creative phase will concentrate on these functions. Worth is defined as 'the lowest overall cost to perform a function without regard to criteria or codes'. Having established the worth and the cost the value index can be calculated. The formula is: value = worth/cost. The benchmark is to achieve a ratio of 1.

The FAST diagram illustrated in Figure 2.15 is characterised by the following:

• The vertical 'scope line', which separates and identifies the highest level function – the task or mission – from the basic and

Table 2.3 Example of cost allocation to function

Element: 2G Internal Walls and Partitions	Elemental cost: £430 283
Function	*Cost (£)*
3.1 Indentify patients	12 000
3.2 Maintain records	30 000
3.3 Study diagnosis	25 900
3.4 Increase availability	6 900
3.5 Maintain hygiene	7 000
4.1 Refer patient	45 000
4.2 Treat patient	56 000
4.3 Process records	40 000
4.4 Circulate people	60 889
5.1 Counsel patient	26 605
5.2 Reduce stress	4 989
5.3 Protect privacy	38 000
6.1 Comfort patient	34 000
6.2 Appear professional	43 000
	£430 283

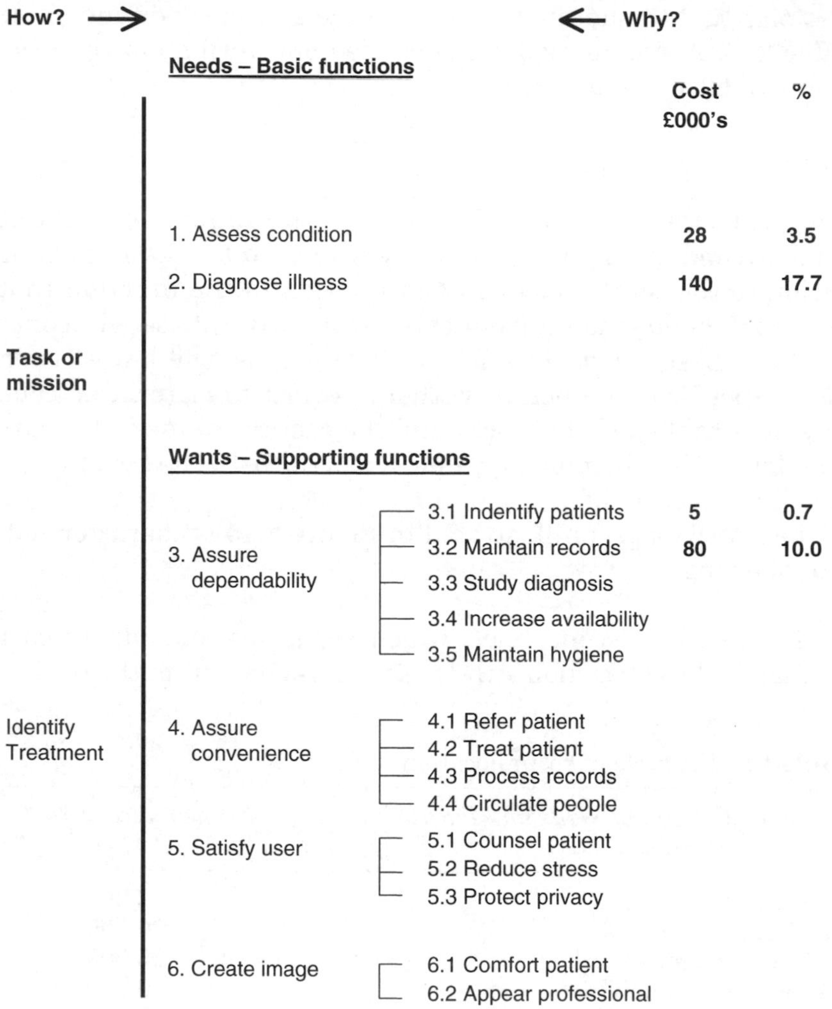

How? ➡   ⬅ Why?

Needs – Basic functions

		Cost £000's	%
1. Assess condition		28	3.5
2. Diagnose illness		140	17.7

Task or mission

Wants – Supporting functions

3. Assure dependability	3.1 Indentify patients	5	0.7
	3.2 Maintain records	80	10.0
	3.3 Study diagnosis		
	3.4 Increase availability		
	3.5 Maintain hygiene		

Identify Treatment

4. Assure convenience
- 4.1 Refer patient
- 4.2 Treat patient
- 4.3 Process records
- 4.4 Circulate people

5. Satisfy user
- 5.1 Counsel patient
- 5.2 Reduce stress
- 5.3 Protect privacy

6. Create image
- 6.1 Comfort patient
- 6.2 Appear professional

Figure 2.15 Functional Analysis System Technique (FAST) diagram for Cancer Treatment and Research Clinic.

supporting functions. It is pivotal to the success of a functional analysis diagram that this definition accurately reflects the mission of the project.

- The division of functions into Needs or Basic functions – with these functions, the project will not meet client requirements and Wants or Supporting functions (usually divided into the four groups previously discussed). The project could still meet the client's functional requirements if these wants are not met or included.
- The use of verb/noun combinations to describe functions.

- Reading the diagram from the left and asking the question 'how is the function fulfilled?' provides the solution.
- Reading the diagram from the right and asking the question 'why?' identifies the need for a particular function.
- The right-hand side of the diagram allows the opportunity to allocate the cost of fulfilling the functions in terms of cost and percentage of total cost.

Therefore, the FAST diagram (Figure 2.15) clearly shows the required identified functions of the project, together with the cost of providing those functions. What now follows is the meat of the workshop – a creative session that relies on good classic brainstorming of ideas, a process that has been compared by those who have experienced it to a group encounter session, the aim of which is to seek alternatives. The discussion may be structured or unstructured – Larry Miles was quoted as saying that the best atmosphere to conduct a study was one laced with cigarette smoke and Bourbon, but in these more politically correct times these aids to creativity are seldom employed. The rules are simple. Nobody is allowed to say, 'That won't work'. Anybody can come up with a crazy idea. These sessions can generate hundreds of ideas, of which perhaps 50 will be studied further in the workshop's Evaluation phase. Those ideas will be revisited and some discussion will take place as to their practicality and value to the client. Every project will have a different agenda. The best of the recommendations are then fully developed by the team, typically on day 4 of the workshop, and studies are carried out into costs and through life costs of a proposed change before presentation to the client on the final day. It is an unfortunate fact of life of the classic 5-day workshop that the team member tasked with costing the recommendations has to work into the night on the penultimate day. Finally, a draft report is approved and a final report is written by the team leader. In addition to the above procedures risk assessment can or, as is thought in some circles, should be introduced into the process. As the value analysts go through and develop value recommendations they can be asked to identify risks associated with those recommendations, which can either be quantitative or qualitative. And if brainstorming sounds just a little esoteric to the quantity surveying psyche, take heart; the results of a value engineering workshop usually produce tangible results that clearly set out the costs and recommendations in a very precise format (Figure 2.16).

The question is often asked are there projects that are beyond value management? The answer is most certainly – yes. There are

PROJECT Cancer Treatment and Research Clinic	VALUE ENGINEERING PROPOSAL	
PROPOSAL Eliminate return duct to ventilation system	DATE	
	ITEM　No　　H14	

ORIGINAL PROPOSAL:

Each room has a return grille and ductwork connecting back to a return fan.

PROPOSED CHANGE:

Eliminate duct return system on individual floors and provide an above ceiling return plenum.

ADVANTAGES:

More available ceiling space

Balancing of return system is simplified

DISADVANTAGES:

Plenum rated cable, tubing and pipe required

May be acoustic transmission problems in walls

COST SUMMARY	INITIAL COST	OPERATION AND MAINTENANCE COST – 15 Years		TOTAL LIFE CYCLE COST
		PER　ANNUM	LIFE CYCLE – PV @ 6%	
ORIGINAL PROPOSAL	£149 450	£4 000	£38 848	£188 298
VE PROPOSAL	£86 000	£2 000	£19 424	£105 424

Figure 2.16 Costs and recommendations of the value engineering proposal.

many high-profile examples that flaunt the drive to lean construction and these mainly fall into the category of projects for which making a statement either commercially, politically or otherwise is their primary, their highest order function. Flyvbjerg in his book *Megaprojects and Risk: An Anatomy of Ambition* cites several examples of international megaprojects that have developed their own unstoppable momentum.

Concurrent engineering

Concurrent or simultaneous engineering deals primarily with the detail design stage. The term refers to an improved design process characterised by rigorous upfront functional analysis, incorporating the constraints of subsequent phases into the conceptual phase, in contrast to the traditional sequential design process.

There can be little argument that the greatest improvements through the implementation of supply chain management techniques, including JIT, have been achieved in the industries where high volume, repetitive manufacture is the norm. In fact, some commentators have suggested that truly successful implementation can only be attained in Japan because of the unique cultural factors and unique employee/employer relationships that exist there. Whether supply chain management techniques are suitable for every project and there is no suggestion by the author that it is, is still a matter of opinion. It is not a panacea for problems of every project from small one-off projects to large multi-million pound infrastructure schemes. However, even if only some of the processes described above are introduced, then if nothing else the focus on delivering client-orientated building solutions that function efficiently will have started. From other industries that have adopted value chain management comes the warning – be patient. Lean production is not a quick fix and major changes in mindset and skills take time: at least 1–2 years for basic understanding, another 3–4 years for training and 2–4 years to achieve sustaining skills and behaviours. One of the most powerful tools in the successful operation of supply chain management are various commercial web-enabled management tools that permit real-time communication between the various supply chain members – this aspect will be discussed further in Chapter 5. To some within the industry, a logical step from the wide scale use of supply chain management techniques is prime contracting, or as it is sometimes known, the 'one-stop shop' approach to procurement, and is discussed in the following chapter.

Conclusion

The government's contract guidance for public construction, announced in May 2000, requires all central government clients to limit their procurement strategies for the delivery of new works from 1 June 2000, and for refurbishment and maintenance contracts from 1 June 2002, to PPPs, design and construct and prime contracting, unless it can be clearly shown that a traditional, non-integrated strategy offers best value for money. In announcing this policy, the Chief Secretary to the Treasury, stated that:

> The Achieving Excellence initiative, launched last year to improve significant Government clients' performance, made it clear that we will focus on interacting with suppliers in the future through integrated supply chains working co-operatively.

In the private sector also the M4i and Constructing Excellence are continuing to drive forward the message about the integration of supply chain management.

Bibliography

Ashworth, A. (1994). *Contractual Procedures in the Construction Industry, Edition* 2, Longman.

Auditor General (2001). *Modernising Construction*, HMSO.

Berger, M. (1994). *Well Connected*, Japan Scope.

Bertelsen, S. and Koskela, L. (2002). *Managing the Three Aspects of Production in Construction*. Proceedings 9th Annual Conference of the International Group for Lean Construction.

Booker, C. (2003). Extravagance in Edinburgh, Sunday Telegraph, April 22, p. 24.

Building (2002). A bit of strategic thinking, 13 September, p. 3.

Building Magazine (2001). The end of a dynasty, 5 October.

CECA (2002). *Supply Chain Relationships*, Civil Engineering Contractors Association, October.

Cutts, R. (1992). Capitalism in Japan: Cartels and Keiretsu, Harvard Business Review.

DTI (2001). *Construction Statistics Annual*, HMSO.

Flybjerg, B. *et al.* (2003). *Megaprojects and Risk: An Anatomy of Ambition*, Cambridge University Press.

Green, S.D. (1999). The Future of Lean Construction: A Brave New World? Proceeding 7th International Lean Thinking Congress.

Green, S.D. (1999). The Dark Side of Lean Construction: Exploitation and Ideology. Proceedings 7th International Lean Thinking Congress.

Harvey, J. (2000). *Urban Land Economics*, fifth edition, MacMillan Press.

Howell, G.A. (1999). *What is Lean Construction?* Proceedings 7th International Conference for Lean Construction, Berkeley, CA, July 26–28.

Howell, G.L. and Macomber, H. (2002). *A Guide for New Users of the Last Planner System – Nine Steps for Success*, Lean Project Consulting Inc.

Koskela, L. (1992). *The Application of the New Production Philosophy to Construction*, Technical Report No. 72, CIFE, Stanford University, CA.

Richards, M. (2003). Budget? Fudge it! *Building Magazine*, 9 May, pp. 49–53.

Simon, P. *et al.* (1997). *Project Risk Analysis and Management*. The Association of Project Management.

The Economist (1991). Inside the charmed circle, Japan's industrial structure, 5 January.

The Financial Times (1994). Mitsubishi's extended family, 30 November.

Whingeing (1991). Japanese-American Trade, The Economist, 18 May.

Winch, G. *et al.* (eds) (2003). *Building Research and Information*, vol 31. Spon Press.

Womack, J. P. *et al.* (1990). *The Machine that Changed the World*, Harper Perennial.

Websites

www.clientsuccess.org.uk/
www.ensembl.org/
www.leanconstruction.org/

3
Managing value.
Part 2: Integrated project delivery

Introduction

Following on from the previous chapter where the advantages of integrated project teams and supply chains were discussed, this chapter reviews some other techniques currently in use by quantity surveyors to ensure integrated project delivery including:

- Partnering and alliancing
- Prime contracting
- Sustainability/Whole-life costs.

Partnering and alliancing

The terms 'partnering' and 'alliancing' do not have the same legal connotations as partnership or joint venture and therefore there has been a tendency, particularly in the construction industry, to apply them rather loosely to a whole range of situations many of which clearly have nothing to do with the true ethos of partnering or alliances, which is to be regretted.

Partnering is viewed by many in the construction industry to be a radical departure from the traditional approach, not only to procurement, but also managing supply chain relationships within the construction industry. Generally received positively by the construction industry, the use of the term 'partnering' in connection with procurement emerged in the USA in the early 1980s. Overall, partnering has received a mixed press in terms of improved performance. For instance, to some observers, partnering has been seen to try to impose a culture of win/win over the top of a commercial framework which remains inherently win/lose. The verbal commitments during the partnering process, even if genuine at the time, are not enough to withstand the stress imposed by gross misalignment of commercial interests. Some critics (Howell *et al.*, 2002), go further and describe partnering as nothing more than a programmatic 'band aid' on the current construction system whose fundamental weaknesses gave rise to its adoption. Research by Ng *et al.* (2002) also seems to confirm the failure on the part of clients to genuinely pursue win/win outcomes. Other studies conducted across other industries, where partnering and supply chain management is common suggest that despite the best intentions, clients easily revert to cost-based criteria, rather than value for money and that rarely do the supply chain members share a common purpose.

When the US process industry became concerned about the lack of competitiveness of its engineering contractors, it established a client-based round table that concluded that long-term relationships with a few contractors should result in cost savings. Partnering therefore originated as a technique for reducing the costs of commercial contract disputes with improving communications, process issues and relationships as the principal focus. The term 'partnering' rather than partnerships was chosen for this procurement strategy because of the legal implications of partnerships. However, to some observers partnering is still an ambiguous term to which at least half a dozen different perspectives may be applied.

By having a smaller number of firms to work with, the client gains considerable benefits. The partnering organisations may gain greater experience of the clients, needs, use techniques such as standardisation of components and processes, bulk purchasing and achieve continuous improvement. Sharing lessons between organisations and applying new ideas ensures the client is getting best practice. Other organisations using partnering have found:

- British Airport Authority has reduced the cost paid for steel from £4000/tonne to £1000/tonne through standardisation
- ASDA, Tesco and Sainsburys are now building supermarkets in 17 weeks rather than 35, with ASDA building four for the price of three
- Jaguar can build a car for the same cost as they used to pay for the components

- Petrochemical companies have used partnering processes for many years, which has shown significant benefits including cost time, quality and health and safety.

Some enthusiasts would claim that partnering is a panacea for the construction industry's ills; pragmatists recognise that it is only one of many solutions that may be appropriate for some, but not all, situations. The Construction Industry Board's report *Partnering in the Team* (1997) states that:

> It is acknowledged that partnering is not an appropriate procurement strategy for all construction projects or programmes ... Partnering succeeds best ... where ... the project or programme is high value and high risk [and] the contractor's interest is fuelled by the prospect of a high value/high attractiveness account core to their business.

There is also evidence, according to the Civil Engineering Contractors Association, that in the civil engineering industry, it seems that some clients have entered into partnering arrangements with contractors, or have let framework agreements without fully appreciating all that is required for these arrangements to be successful in terms of delivering better value. In particular, some clients seem not to have looked much beyond the subsidiary objective of these arrangements, which is to secure savings in costs of procurement and contract administration. There is evidence of what has been labelled 'institutional pressure', that is to say clients and contractors feeling that they must move in the directions in which the Latham and Egan reports are pointing them, but there is a danger that they will begin to move on the basis of insufficient knowledge and understanding of what is required. According to Wood (2005) numerous authors have tried to analyse the critical success factors for successful partnering relationships, however despite some differences from various studies, the assertion made by Bennett and Jayes (1997) that true partnering relies on co-operation and teamwork, openness and honesty, trust, equity and equality is still appropriate.

Managing the supply chain

The preferred approach to managing the supply chain is partnering. It has been described as welding the links of the supply chain together. Although the term 'partnering' is relatively new, having been adopted in various guises within the UK construction industry since the late 1980s, this is not the case with the relationship itself. Some contractors had been practising what they might term 'collaborative contracting' for many years before the term partnering was adopted with respect to a formal arrangement – for example, Bovis' relationship with Marks and Spencer.

Essentially, partnering enables organisations to develop collaborative relationships either for one-off projects (project-specific) or as long-term associations (strategic partnering). The process is used as a tool to improve performance, and may apply to two organisations (e.g. a client and a design and build contractor) or to a number of organisations within a formal or informal agreement (e.g. consultants, contractors, subcontractors, suppliers, manufacturers, etc., with or without client participation). The partnering process is formalised within a relationship that might be defined within a charter or a contractual agreement.

Partnering is a structured management approach to facilitate teamwork across contractual boundaries that helps people to work together effectively in order to satisfy their organisations' (and perhaps their own) objectives. It is seen by many as a means of avoiding risks and conflict. There is no single model partnering arrangement; it is an approach that is essentially flexible, and needs to be tailored to suit specific circumstances.

Words and phrases selected from numerous definitions or descriptions of partnering relate either to the process (such as avoid waste and conflict, effectively communicate, co-operate/collaborate, work together/teamworking, integrated, co-ordinated, continuous evaluation, avoid/minimise/share risk) or to the desired outcomes (such as achieve mutual/common objectives, achieve continuous, measurable improvement, maximise/share benefits/rewards). The National Economic Development Office report Partnering: Contracting without Conflict (NEDO, 1991) adapted the following definition, from one which was originally produced by the United States Construction Industry Institute:

> ... partnering is a contractual arrangement between a client and his chosen contractor which has a term of a given number of years rather than the duration of a specific project. Thus ... contractors may be responsible for a number of projects...

Bennet and Jayes (1995) streamlined the definition and identified that it can be project specific. They also added 'method of problem resolution' to the process, and 'measurable' to continuous improvement. Their definition also makes reference to it being a 'management approach' rather than NEDO's reference to a contractual arrangement. In the report, Trusting the Team, partnering is defined as follows:

> Partnering is a management approach used by two or more organisations to achieve specific business objectives by maximising the effectiveness of each participant's resources. The approach is based on mutual objectives, an agreed method of problem resolution and an active search for continuous measurable improvements. Partnering can be based on a single project (project partnering) but greater benefits are available when it is based on a long-term commitment (strategic partnering).

This reflects how the perception of partnering has developed and changed in a relatively short period of time. The understanding of the process had further developed by 1997, when Bennet and Jayes, with the Partnering Task Force of Reading Construction Forum, published The Seven Pillars of Partnering. This report defined partnering as:

> ... a set of strategic actions which embody the mutual objectives of a number of firms achieved by co-operative decision-making aimed at using feedback to continuously improve their joint performance.

Here, problem-solving is subsumed within decision-making in what they term second generation partnering. This approach, which recognises that more significant benefits can be achieved through long-term relationships throughout the supply chain, 'begins with a strategic decision to co-operate by a client and a group of consultants, contractors and specialists engaged in an ongoing series of projects'. The Reading Construction Forum's research suggests that a third generation is emerging where organisations form an alliance to utilise fully the expertise within the supply chain in order to provide comprehensive packages, similar to the PFI, where clients wish to outsource construction and perhaps facilities management enabling them to concentrate on their core business.

Overview – a client's perspective

Government-led initiatives have repeatedly expressed the wish to see partnering become the norm in the hope that it will promote a new way of working. It is clear that public sector clients (including local authorities) are being directed towards procurement strategies that are based on integration and collaboration. The National Audit Office's report, Modernising Construction, gave support for public sector clients in promoting innovation and good practice, encouraging the industry as a whole and its clients to:

- Select contractors on the basis of value for money
- Develop close working relationships between clients and the entire supply chain
- Integrate the entire supply chain, including clients, professional advisors, designers, contractors, subcontractors and suppliers.

In the private sector, many major and influential clients across all sectors have been adopting partnering in response to the proven long-term benefits that can be achieved through this approach. Companies such as Sainsbury's, BAA and Esso are reported to have reached savings of up to 40 per cent on costs and 70 per cent on time by using partnering approaches. There is, however, evidence that small, occasional clients have little to gain from the process.

Overview – a contractor's and consultant's perspective

Main contractors publicly at least, appear to have enthusiastically embraced the partnering concept, without which much of the work available from the major clients is not accessible. General and specialist subcontractors, suppliers and manufacturers may be involved through partnering within a larger supply chain, but many claim that the only benefit for them is assured workload, although this comes at a price – lower profit margins, for example. Nevertheless, it is interesting to note the growth of networking events/marketplaces that offer manufacturers the opportunity to forge new relationships by providing a consultation service on their stands to promote their design and problem-solving skills, rather than selling products. These collaborative, solutions-driven events have been enthusiastically supported by major materials manufacturers, who recognise the contribution they are already making to design through early involvement in partnering arrangements.

Many large client organisations now have framework agreements with consultants as well as contractors, covering periods of time and a series of projects, and of course many consultants have been 'preferred' firms of regular clients for many years. Framework agreements often do little more than formalise long-term relationships, although some clients are becoming more demanding of their consultants in these agreements, resulting in firms being dropped and others refusing to sign up (see Chapter 7). The main attraction for consultants, large or small, is undoubtedly the security of workload offered by long-term arrangements, but this may be at a price – financial or otherwise. Many consultants, particularly architects, have formed strong relationships with contractors to compete for design and build projects, which have been increasingly attracting many clients for some years. Some of these arrangements are now developing into the core of prime contracting alliances which are discussed later in this chapter. Besides security of workload, attractions of partnering for consultants might include the satisfaction and reputation gained from being associated with successful projects or high-profile clients. Theoretically, greater profits are achievable through sharing in savings; however, there is a lack of hard evidence that consultants benefit from this. Nevertheless, it is likely that, where consultants partner with contractors in, say, a design and build or prime contracting project, the consultants may well share some of the savings awarded to the contractor. Project-specific partnering would not appear to offer many benefits for consultants.

There are numerous examples to be found on construction-related websites, such as www.constructingexcellence.org.uk that illustrate partnering success stories, however anecdotal evidence suggests that perhaps partnering is littered with potential pitfalls. An RICS research report Beyond Partnering: Towards a New Approach in Project Management, published in 2005, aimed to examine how the barriers to partnering success, culture and the economic reality of supply chain relationships, are being addressed in practice. In addition the study attempted to assess the actual benefits that accrue from partnering and whether there is any real change in construction industry practice. The RICS study used a series of semi-structured interviews with a relatively small sample of senior figures within 10 major construction clients, including large retailers and utilities organisations with a combined annual construction spend of approximately £2000 million. The client sample procured between 50 and 100 per cent of their total expenditure using partnering arrangements. In addition 10 national contracting organisations were selected with a combined annual turnover of approximately £4350 million of which £1500 million is delivered through partnering arrangements.

Respondents were unanimous that to be successful partnering arrangements, which in some cases included alliancing which is discussed later, require a culture change within the industry on both supply and demand sides. An inherent lack of trust manifested itself as a threat to successful partnering which was identified as lack of openness and honesty of clients and a 'Luddite' culture within contractor's organisations, resulting in little change of practice at site level. Interestingly, the survey heard 'that it is quantity surveyors who find it most difficult to adapt, since they are used to problems being addressed in a contractual and confrontational manner, rather than by people communicating in order to find solutions'. On economic issues clients admit to obtaining commercial leverage over their supply chains and there is evidence that the subcontractor squeeze described previously in Chapter 2, is still alive and well in supply chains. For contractors the continuity of working repeatedly for the same clients is also thought to provide a number of benefits for contracting organisations, although whether a contractor's profit margin increases on partnering projects is unclear. Both supply and demand sides agreed that partnering provided a more rewarding environment in which to operate.

Key success factors

Simply adopting a policy of partnering with and within the supply chain will not itself ensure success. Partnering is not an easy option; a number of prerequisites, or key success factors, need to be taken on board. Some of the following are desirable for project partnering; all are essential for successful strategic partnering:

- There needs to be a commitment at all levels within an organisation to make the project or programme of work a success, which means a commitment to working together with others to ensure a successful outcome for all participants (win–win situation). Returns will not be immediate; a willingness to make an early (and perhaps not insignificant) investment in time and effort to build the team is essential.

- Partners must have confidence in each other's organisations, and each organisation needs to have confidence in its own team, which means careful selection of the people involved. Participants need to have a clear understanding and commitment to the teamworking culture. Partners should be chosen on the basis of the ability to offer best value for money and not on lowest price; their ability to innovate and offer effective solutions should also be considered.

Clients should normally select their partners from competitive bids based on carefully set criteria aimed at getting best value for money. This initial competition should have an open and known prequalification system for bidders.

- Partners need collectively to agree the objectives of the arrangement/project/programme of work and ensure alignment/compatibility of goals. This will require early involvement of the entire team to ensure a win–win situation for all. It is this agreement that should drive the relationship, not the contract(s). The agenda must be of mutual interest with a focus on the customer; it must therefore be quality/value driven.
- To satisfy the relationship's agenda, there needs to be clarity from the client and continued client involvement. It is essential to define clearly the responsibilities of all participants within an integrated process. There can be no weak links. People, without regard to affiliation, must be brought together into integrated teams with streamlined supply chain management. There needs to be a willingness to be flexible and adopt new ideas and different ways of doing things, for example, different operational methodologies, different administrative procedures, different payment methods, different payment procedures, etc.
- Sharing is important. All players should share in success in line with their contribution to the value added process (which will often be difficult to assess). There also needs to be a sharing of information, which requires open-book accounting and open, flexible communication between organisations/teams/people. Responsibility for risks must be allocated clearly and fairly, but there must be a collective responsibility for problems and an openness and willingness to accept and share mistakes. This requires a departure from the finger-pointing, blame culture to an acceptance that getting things wrong results in a lose–lose,

rather than a win–win, scenario. Adoption of such openness and sharing requires trust.

- It is important that all partnering arrangements incorporate effective methods of measuring performance. It has been identified that partnering should strive for continuous improvement, and this must be measurable to ascertain whether or not the process is effective. It is essential therefore that agreed measurable (and achievable) targets for productivity improvement are set, and clear, easily understood measurement systems are adopted to evaluate efficiency with respect to time and cost. Similarly, ways of measuring improvements in quality must be adopted as part of the quality assurance procedures. Benchmarking, perhaps including use of the industry's key performance indicators (KPIs), should be adopted and, to enable organisations to share in savings/increased benefits, effective incentive schemes need to be in place. In long-term agreements, annual reviews should take place to reset objectives/targets. Clients with major programmes of work should not put all their eggs in one basket; this will enable them to benchmark their different partners against each other. These clients may periodically test the market by limiting the time of agreements to enable new bids to be made; however, it is important that the duration of agreements is of reasonable length to assure bidders of predictable turnover (subject to satisfactory performance). This will, of course, require clients to be able to assure continuity (or at least predictability) of workload.
- There will be times when partners don't agree, and it is therefore important that agreed non-adversarial conflict resolution procedures are in place to resolve problems within the relationship. The principle of trying to resolve disputes at the lowest possible level should normally be adopted to save time and cost.
- Education and training is needed to ensure an understanding of partnering philosophy. It is important that, regardless of how well-versed participants are in the philosophy and procedures, teambuilding takes place at commencement of the relationship. Good teambuilding and development of efficient teamworking can be enhanced by adopting commonsense procedures such as sharing of office space and sharing of information through computing networks/intranets, etc. Successful teambuilding should result in the development of trust.

Trust is generally regarded as being crucial to the success of partnering; indeed it has been described as the cornerstone of a successful partnering relationship. Blois (1999) argues that only individuals can trust, and consequently trust between organisations (which are, after all, only collections of people) means a lot of people needing to trust a lot of other people. This makes relationships vulnerable to human fickleness. Careful selection of the correct individuals by each organisation is therefore crucial to success. These individuals, particularly those at the interface with partners, must fully accept the philosophy of partnering and be committed to the success of the arrangement. Changes in personnel or attitudes of key people can render relationships fragile; it is therefore very important to invest in trust building activities amongst everyone involved with integrated teams so that the collective trust in the team as a whole can withstand breakdown in individual relationships. The building of trust, which is essential for co-operative activity, takes time, effort and patience, particularly for those with experience of the traditional way of doing business. Where trust has been established, partners can be relied upon.

Reliance on others to do what is expected is necessary in any arrangement, including conventional procurement strategies, where a contract is used to encourage/enforce it. This extra ingredient, which stems from trust, is what is being sought within partnering arrangements. There will be times when organisations in a partnering arrangement are tempted not to act in a mutually acceptable way. This may be for short-term gain, or market changes may result in changes in attitude. In such times, many organisations may be tempted to break faith and break ranks. Only when trust has been fully developed will commitment to seeing things through be expected of partners.

Opportunities for quantity surveyors

Collaborative integrated procurement offers opportunities for quantity surveyors, including:

- *Acting as an independent client advisor*. Many clients will still look to their quantity surveyor for independent advice. This raises the question, 'where is the trust within the relationship if external advice is still needed?' Many clients will still feel that they need advice of someone without an axe to grind – for example, appointing an external quantity surveyor and audit team to ensure that its strategic partners perform. Services provided might include assessment of target costs, development of incentive schemes, measurement of performance, auditing, etc.

- *Participating as a partner in an alliance.* Quantity surveyors able to demonstrate that they have the skills and ingenuity to add value will be welcomed by most alliances. Imaginative teams will consider numerous solutions – who better to evaluate the alternatives than the quantity surveyor?
- *Leading an integrated supply chain.* Many quantity surveyors have become successful project managers. There is no reason therefore that they cannot manage a supply chain. With appropriate financial resources, a quantity surveying practice can act as a prime contractor.
- *Acting as a partnering advisor within PPC 2000 contracts.* The described role would seem to fit the quantity surveyor with partnering experience. This is a key role and would suit a quantity surveyor who can demonstrate a collaborative rather than adversarial attitude.

The move towards clients partnering with integrated supply chains offers significant opportunities for consultants wishing to join alliances to share in the potential rewards. If the industry does become less adversarial, as is hoped, quantity surveyors will welcome it. They will then be able to concentrate on what they do best – adding value for clients, which coincides with the purpose of partnering!

Alliancing

As with partnering, alliancing can be catagorised as follows:

Strategic alliances can be described as two or more firms that collaborate to pursue mutually compatible goals that would be difficult to achieve alone. The firms remain independent following the formation of the alliance. Alliancing should not be confused with mergers or acquisitions.

A project alliance is where a client forms an alliance with one or more service providers: designers, contractors, supplier, etc. for a specific project and this section will continue to concentrate on this aspect of alliancing.

The principal features of a project alliance are as follows:

- The project is governed by a project alliance board, that is composed from all parties to the alliance that have equal representation on the board. One outcome of this is that the client has to divulge to the other board members far more information than would, under other forms of procurement, be deemed to be prudent.
- The day-to-day management of the project is handled by an integrated project management team drawn from the expertise within the various parties on the basis of the best person for the job.
- There is a commitment to settle disputes without recourse to litigation except in the circumstance of wilful default.
- Reimbursement to the non-client parties is by way of 100 per cent open book accounting based on:

  1. 100 per cent of expenditure including project overheads: Each non-client participant is reimbursed the actual costs incurred on the project, including costs associated with re-works. However, reimbursement under this heading must not include any hidden contributions to corporate overheads or profit. All project transactions and costings are 100 per cent open book and subject to audit.
  2. A fixed lump sum to cover corporate overheads and a fee to cover profit margin: This is the fee for providing services to the alliance, usually shown as a percentage based on 'business as usual'. The fee should represent the normal return for providing the particular service.
  3. Pain/gain mechanism with pre–agreed targets: The incentive to generate the best project results lies in the concept of reward, which is performance based.

A fundamental principle of alliances is the acceptance on the part of all the members of a share of losses, should they arise, as well as a share in rewards of the project. Risk: Reward should be linked to project outcomes which add to or detract from the value to the client. In practice, there will be a limit to the losses that any of the alliance members, other than the client, will be willing to accept, if the project turns out badly. Unless there are good reasons to the contrary it may be expected that the alliance will take 50 per cent of the risk and the owner/client the remaining 50 per cent. The sharing of pain:gain is generally based on objectively measurable outcomes in key performance areas, such as:

- Time of delivery
- Safety
- Environmental compliance
- Industrial and community relations.

Performance-based remuneration ensures that some of the contractor's remuneration, the profit margin referred to in point 2, is at risk unless it achieves the indicators.

The major differences between alliancing and project partnering are outlined in Table 3.1.

For example, in project partnering one supplier may sink or swim without necessarily affecting the business position of the other suppliers. One entity may make a profit, while the other entity makes a financial loss. However, with alliancing there is a joint rather

Table 3.1 Differences between alliancing and project partnering

	Partnering	*Alliances*
The form of the undertaking	Core group with no legal responsibilities. Binding/non-binding charters used in 65% of partnering arrangements	Quasi joint venture operating at one level as a single company
The selection process	Prime contractor responsible for choice of supply chain partners. Project can commence while selection continues	Rigorous selection process Alliance agreement not concluded until all members appointed
The management structure	By prime contractor Partnering advisor	Alliance board
Risk and reward mechanisms	Partners losses not shared by other members of the supply chain	Losses by one alliance member shared by other members

than a shared loss, therefore if one alliance party underperforms then all the parties are at risk of losing.

Therefore, alliance members form a quasi joint venture, because they operate at one level as a single company; however, they do not merge their companies in any legal sense. They remain independent but they must work with each other in order to meet the Key Performance Indicators (KPIs) (see Chapter 1), to realise risk and reward. Therefore if the project fails to meet agreed project KPIs then all alliance members share the loss.

There are some significant legal and financial aspects that need to be put in place in an alliance agreement, but while important, it is the behaviour of the parties which determines whether an alliance will be successful. Choose your alliance members carefully!

Therefore, given the operational criteria of an alliance, it is vitally important that members of the alliance are selected against rigorous criteria. These criteria, usually demonstrated by reference to previous projects undertaken by the prospective alliance members, vary from project to project but as a guide could be:

- Demonstrated ability to complete the full scope of works being undertaken from the technical, financial and managerial perspectives
- Re-engineer project capital and operating costs without sacrificing quality
- Achieve outstanding quality with an outstanding track record
- Innovate and deliver outstanding design and construction outcomes
- Demonstrate safety performance
- Demonstrate conversance with sustainability issues
- Work as a member of an alliance with a commitment to non-adversarial culture and change direction quickly if required
- Have a joint view on what the risks are and how they will be managed.

Establishing a target cost and dealing with variations

One of the key tasks in the establishment of the alliance board is the development of the target cost – it is not possible to do this before this point. The target cost is intended to be the best estimate of what the integrated teams think the project will cost. It is important that all alliance members feel comfortable with target cost which should allow for inherent uncertainties consistent with the state of knowledge at the time of preparation. Under a conventional contract, the tender sum is just the starting point with subsequent variations and claims resulting in a substantially higher final account cost to the client. In a project alliance the target cost must allow for all matters that would normally be the subject of a variation under a conventional JCT (05) contract. The alliance members collectively assume responsibility for all sorts of risks that are normally retained by the client under a traditional approach for example:

- Design changes
- Late delivery by suppliers
- Inclement weather

and reasonable provision has to be made within the target cost for such items.

The circumstances under which variations arise are limited. Generally, normal changes due to design development are not considered. Changes that could give rise to variations are:

- Significant increases or decreases in the scope of work, e.g. adding in new buildings, parts of buildings or facilities
- Fundamental changes in the performance parameters.

If the alliance members agree that a variation is declared then members fees as well as the target cost are adjusted.

One of the benefits of the alliance approach is that in the traditional approach to procurement each contractor, as well as the client, will include a sum for contingencies for unforeseen circumstances, whereas in an alliance profit and overheads are agreed in advance and the contingency allowance can be reduced and managed collectively. Although there have been some high-profile success stories with alliances, in general failure rates for alliances are high at around 50–60 per cent, a salutary lesson for those about to dash head long into this form of procurement. Reasons for alliance failure, and conversely critical success factors have been identified as:

- Poor partner selection
- Mismatch of organisational structures or culture

- Poor systems for information sharing
- Lack of trust
- Lack of commitment.

And finally, the small print

In this culture of mutual trust and openness, is there a need for lawyers? The answer is of course – Yes! Lawyers have an important role to play in ensuring that the intention of the parties is enshrined in a properly structured and legally effective alliance agreement. Some of the more notable features of typical alliance agreements that set them apart from standard forms of contract include:

- Collective performance obligations
- Details of pain/gain conditions together with statement of extent of liabilities
- Details of audit arrangement for each alliance member.

In April 2002, Masons, a leading law firm specialising in construction matters, published a report entitled Partnering, the Industry Speaks, which contained the results of a survey of over 1000 participants within the construction industry, ranging from contractors, subcontractors and consultants. Of the organisations questioned, 79 per cent confirmed that they had been engaged in partnering, to some extent, but only 6 per cent of organisations had more than 50 per cent of their workload based on partnering arrangements. When questioned about the potential benefits of partnering the respondents proposed:

- Avoidance of disputes / blame culture
- Improved profitability
- Mutual objectives rather than mutual gain
- Greater client satisfaction.

Partnering has been described as a process whereby the parties to a traditional risk transfer form of contract, i.e. the client, the contractor and the supply chain, commit to work together with enhanced communications, in a spirit of mutual trust and respect towards the achievement of shared objectives. As with alliancing, there are two approaches, strategic and project partnering. The differences are as follows:

- Strategic partnering is concerned with a range of work, often unspecified at the time that the contract is made, over a period of time. The motivations are to achieve consistency and predictability of workload, to take out waste and to achieve continuous improvement through experience and learning.
- Project partnering is much more focused on a single project.

The three essential elements of partnering are said to be that the individuals and firms involved in a construction project should agree to:

1. *Mutual objectives*. Mutual objectives to which all parties are fully committed. All the parties open up about their own objectives so that they get a better understanding of what each organisation is trying to achieve. Following this a set of mutual objectives can be drawn up, which in the case of project partnering are specific to a project taking into account the individual organisations' objectives, that can form the basis of the partnering charter. Examples of mutual objectives are illustrated in Table 3.2.
2. *Decision-making*. Secondly, partnering requires agreement on how decisions are made, including how any disputes will be resolved. One of the basics of partnering is empowerment and the idea that decisions should be made and problems solved at the lowest levels. In addition, partnering places importance on the establishment of non-contractual ways of quickly resolving such matters.
3. *Continuous improvement*. Finally, the parties must commit to seeking continuous and measurable improvement in their

Table 3.2 Examples of mutual objectives

Objective	How Achieved
1. Improve efficiency	1. Co-operation
2. Cost reduction	2. Continuous improvement
3. Cost certainty	3. Early action on high-risk areas
4. Enhanced value	4. Buildability, value engineering
5. Reasonable profits	5. Predictable progress
6. Reliable product quality	6. Quality assurance / TQM

performance. Key performance indicators are set by the partnering parties who decide among themselves what should be
measured and what target should bc set. Typically the items
measured are:

- Base cost
- Time
- Quality.

There will obviously be many road blocks in trying to introduce a
system such as partnering into a traditional construction organisation and these can be catagorised as cultural and commercial issues.

Cultural issues

Cultural realignment for an industry is difficult to achieve. The approach generally adopted by organisations is to organise partnering
workshops during which the approach is explained. This includes:

- The establishment of *trust* – a vital in the partnering process
- Emphasising the importance of *good communications* between
  partnering parties
- Explaining that many activities are carried out by *teamwork*
  and how this is achieved in practice
- The establishment of a *win/win* approach
- Breaking down traditional management structures and *empow-
  ering* individuals to work in new ways.

Commercial issues

Organisations wishing to be involved in partnering arrangements
should be able to demonstrate capabilities in the following:

- Establishing a *target (base) cost* that is based on open book accounting and incorporating pain/gain schemes. Profit margins
  should be established and ring fenced.
- A track record of using *value engineering* to produce best value
  solution.
- A track record of using *risk management* to identify and understand the possible impact of risk on the project.

- An ability to successfully utilise *benchmarking* to identify how to measure and improve performance.
- The willingness to take a radical look at the way they manage their business and to engage in *business process re-engineering* if necessary to eliminate waste and duplication.

So where do the potential problems and pitfalls lie, or is partnering the simple route to trouble-free contracting? The Masons' survey, Partnering, The Industry Speaks, highlighted the most significant potential pitfalls of partnering as:

- Having *too high expectations*. Parties will have to work hard to overcome amicably and fairly problems and conflicts that arise
- Having *disparate objectives*. Partnering requires the alignment of objectives and an understanding of an interest in each other's success
- Setting *unrealistic risk allocation*. It is essential that parties have a realistic attitude to risk and how to manage it efficiently and effectively
- *No 'buy-in' at all levels*. Full benefits in the procurement process will only be achieved if all levels of the process embrace the partnering approach.

Partnering contracts

A common sentiment expressed by various commentators is that partnering is a non-legal or 'moral' relationship that sits alongside the formal legal/contractual relationship between the parties. Some argue that we need to step away from the traditional methods of procurement and project delivery and if this means scrapping the formal written contract, so be it. However, not everyone is of this view. One school of thought is that it is important to make partnering arrangements legally binding. The contrary view being that partnering is a consensual process and that its purpose is lost if parties are forced to do it. As part of the partnering process, a so-called non-binding partnering charter is sometimes used to lay over a traditional, adversarial form of contract. The charter uses language such as honesty, trust, co-operation and shared objectives. Using this approach parties try to get partnering benefit with the re-assurance that they can return to the old ways if partnering is not successful. Funders often prefer this approach. Regardless of which view is taken, most lawyers do agree that the parties need to spell out in clear terms, whether the partnering arrangements are or are not legally binding. Until relatively recently, a common approach was for the parties to contract in the usual way and then to set about formulating their partnering arrangements. Often this involved senior management having meetings to thrash out a partnering charter or agreement, which might or might not be legally binding on the parties.

In the past 4 or 5 years, specific contracts and contractual amendments have been drafted to implement partnering as part of the formal contract process. These include:

- The ACA Standard Form of Contract for Project Partnering (PPC 2000)
- The New Engineering Contract, 3rd edition (NEC3) has Option X12 set of clauses. This in effect is a Standard Form of Contract underneath a partnering agreement.

In addition, the following two alternatives are available:

- No contract, but a partnering charter
- Standard form of contract (JCT 05) and partnering charter.

Project Partnering Contract 2000 (PPC 2000)

Even though partnering has been described as a state of mind rather than the opportunity to publish another new contract, there is now available a standard form of contract for using with partnering arrangements – Project Partnering Contract 2000 (PPC 2000) (Figure 3.1). PPC 2000 was developed for individual projects that were to be procured using the partnering ethos. By the nature of the contract it is more suitable to a single project rather than an ongoing framework agreement, covering many projects over a period of time, which would need a bespoke contract. PPC 2000 was developed to clarify the contractual relationship where partnerships were being entered into using amended JCT contracts and in many cases this was proving to be confusing and to the JCT losing its' familiarity. Partnering charters, whilst proving valuable, were not a legal basis for an agreement and did not outline the system for non-adversarial approach. Drafted by Trowers and Hamlins, a leading London law firm, PPC was launched in November 2000 by Sir John Egan and published by the Association of Consultant

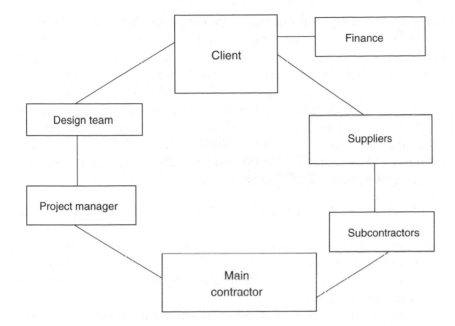

Figure 3.1 PPC 2000 contract.

Architects Ltd (ACA). It includes radical ideas set out in the CIC Partnering Taskforce guide, launched by Nick Raynsford. There is a standard adaptation for use in Scotland.

The key features of PPC 2000 are designed to encourage a team-based approach between client, professional team, contractor and specialists. It is a single multiparty agreement covering all significant aspects of and the entire duration of the procurement process. The single contract approach

- Reduces the danger of gaps or inconsistencies between different documentation
- Sets out clearly the relationships between team members
- Helps move parties away from the confrontational mindset of traditional procurement.

The aim of PPC 2000 is that the team members are in contract much earlier than is traditional and for negotiations to be carried out post-contract. The contract still allows for a partnering charter and also covers consultants' appointments and agreements – one contract for the whole. It is claimed that PPC 2000 is written in straightforward language with the flexibility to allow the team to evolve as the project progresses, also that its flexibility allows for risk to be allocated as appropriate and moved as the project develops. However, because the PPC 2000 is such a departure from the more traditional contracts, it has proved something of a struggle for those who use it for the first time. The contract can be difficult to interpret, as key elements such as risk sharing and key performance indicators are left as headings with the detail to be completed later. To guide new users, in June 2003 a Guide to ACA Project Partnering Contracts PPC 2000 and SPC 2000 was launched.

The key features of PPC 2000 include:

- A team approach with duties of fairness, teamwork and shared financial motivation
- Stated partnering objectives – including innovation, improved efficiency through the use of Key Performance Indicators (KPIs) and completion of the project within an agreed time and to an agreed quality
- A price framework which sets out the contractor's profit, central office overheads and site overheads as well as an agreed maximum price
- A procedure for dispute resolution hierarchy
- Commitment to the most advantageous approach to the analysis and the management of risk
- The ability to take out latent defects and/or project insurance (see Chapter 1).

Some aspects may cause difficulties, for example, Section 3 of the contract gives the client representative the ability to inspect the financial records of any member of the team at any time subject to reasonable notice and access to members' computer networks and data by each member.

The take up of the PPC 2000 contract is thought to be in the region of between £1 billion and £2 billion worth of contracts since its launch in 2000. However welcome the introduction of PPC 2000 and partnering, using this approach as a parallel agenda with traditional standard forms of contract as a fall-back position creates confusion and suggests a lack of commitment.

The SPC 2000 for Specialist Contractors

A further development is the publication of a subcontract to complement PPC 2000 by the ACA. The specialist contract is intended to provide a standard document so that parties entering into PPC 2000 can have back-to-back arrangements with their subcontractors or specialists, to use the contract terminology. The SPC 2000 has the same basic structure as PPC 2000, but includes a specialist agreement to which the specialist terms are appended. The specialist contract endeavours to ensure that the constructor and the specialist work more effectively together than is perceived to be the case under the traditional forms of contract. Unsurprisingly, it prescribes partnering objectives and targets. As with all such provisions, these are a mixture of aspirations and legally enforceable obligations and as such there are question marks over the status, enforceability and implications of the aspirational objectives, particularly in the context of the rights and obligations under the contract. The specialist needs to recognise, however, that although it is working in a partnering arrangement with the constructor, it is also responsible for managing all risks associated with the specialists works unless otherwise provided for. As part of the process, the constructor and the specialist must identify what risks might arise and then share or apportion the risks according to who is more able to manage them.

The NEC3 Partnering Agreement

To compare the PPC 2000 with the NEC3 Partnering Agreement is to compare apples with oranges. Whereas PPC 2000 is a free-standing multiparty contract, governing all the parties' mutual rights and obligations in respect of a particular project, rather than being, as in the case of the NEC3 partnering agreement, an option bolted to a series of biparty contracts, which must each be based on the NEC3 form. When using NEC3 Option X12 each member of the partnering team has its own contract with the client (see Figure 3.2). The NEC3 Partnering Agreement, which by contrast to PPC 2000 is extremely short, acts as a framework for more detailed provisions which must be articulated by the parties themselves in the schedule of partners, or in the document called the partnering information. It is up to the parties to identify the objectives. Further provisions in the Partnering Agreement set out obligations which are an essential condition if those objectives are to be met. For example, attendance at partners and core group meetings, arrangements for joint design development, risk management and liability assessments, value engineering and value management, etc.

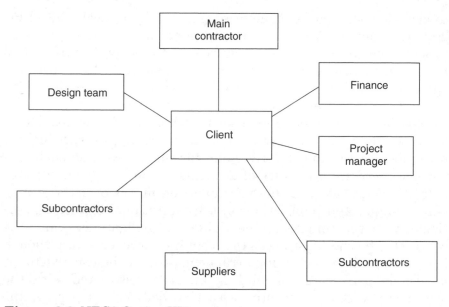

Figure 3.2 NEC3 Option XI2 contract.

Prime contracting

Introduced in the 1990s, prime contracting is a long-term contracting relationship, based on partnering principals and is currently being used by several large public sector agencies, as well as some private sector clients. A prime contractor is defined as an entity that has complete responsibility for the delivery and in some cases, the operation of a built asset and may be either a contractor, in the generally accepted meaning of the term, or a firm of consultants. The prime contractor needs to be an organisation with the ability to bring together all of the parties in the supply chain necessary to meet the client's requirements. There is nothing to prevent a designer, facilities manager, financier or other organisation from acting as a prime contractor. However, by their nature prime contracting arrangements tend to require the prime contractor not only to have access to an integrated supply chain with substantial resources and skills such as project management. To date most prime contractors are in fact large firms of contractors, despite the concerted efforts of many agencies to emphasise the point that this role is not restricted to traditional perceptions of contracting. One of the chief advantages for public sector clients with a vast portfolio of built assets is that prime contracting offers one point of contact/responsibility, instead of a client having to engage separately with a range of different specialists.

As with other forms of procurement based on a long-term partnership, the objective of prime contracting is to achieve better long-term value for money through a number of initiatives such as supply chain management, incentivised payment mechanisms, continuous improvement, economies of scale and partnering. The approach to prime contracting differs from the PFI because the prime contractors' obligations are usually limited to the design and construction of the built asset and the subsequent facilities management; there is no service delivery involved. Also the funding aspects of this approach are much less significant for the prime contractor, in that finance is provided by the public sector client. In some models in current use, prime contractors take responsibility not only for the technical aspects of a project during the construction phase, including design and supply chain management but also for the day-to-day running and management of the project once completed. This may include a contractual liability for the prime contractor to guarantee the whole-life costs of a project over a predetermined period for as much as 30 years.

For the client the major attraction is that the traditional dysfunctional system is replaced by single point responsibility. The prime contractor has the responsibility for:

- Total delivery of the project in line with through life predictions, which can be up to 30 years from the time of delivery
- Subcontractor/Supplier selection – note there are exceptions to this rule, i.e. where a client may, because of its influence or market position be able to procure some items more cheaply than the existing supply chain
- Procurement management
- Planning, programming and cost control
- Design co-ordination and the overall system engineering and testing.

The decision was taken by the MoD to pilot prime contracting because of the difficulties of having to manage a vast estate of diverse properties spread through the UK. For the public sector client, the attractions of prime contracting were as follows:

- Fewer larger long-term contracts
- Shared risk

- Partnering in the supply chain
- Incentivisation of private sector contractors.

The model of PC adopted by the Ministry of Defence is based on the following features:

- The use of an output specification. The key objective is to allow the prime contractor and the supply chain to deliver innovative and more efficient services
- Involvement of the prime contractor at an early stage in the procurement process
- Selection criteria judged on the partnering ethos and collaborative working with greater emphasis on technical and soft issues, such as market awareness, trust and openness, flexibility and understanding the client culture
- Terms and conditions which reflect fitness for purpose and incentivised price and costing regime, open book accounting, supply chain management and dispute resolution
- Integrated project teams.

J. Sainsbury, a private sector company using the prime contracting approach, restricts the inclusion of life cycle costs to an occasional audit. The ever-changing demands of their business requirements do not, at present, fit long term, i.e. 30 years, life cycle cost considerations. As illustrated in Figure 3.3, the client generally has one point of contact, the prime contractor, and this arrangement replaces the traditional situation where a client was confronted with a series of relationships that could be categorised into:

- Contractual and
- Procedural

which resulted in a complex mesh of relationships that seem as if they were designed to provide a smoke screen to hide inefficiency and accountability. It's the equivalent of a new car buyer having to procure the engine, body shell and gear box from independent sources.

Many clients, in both the public and private sectors, do not wish to contract with a whole range of different contractors, designers, suppliers and subcontractors. Clients talk about single-point supply delivering them peace of mind with assurances as to time cost and quality from a robust prime contractor able to manage and harness effectively the value stream outputs from the supply chain. As

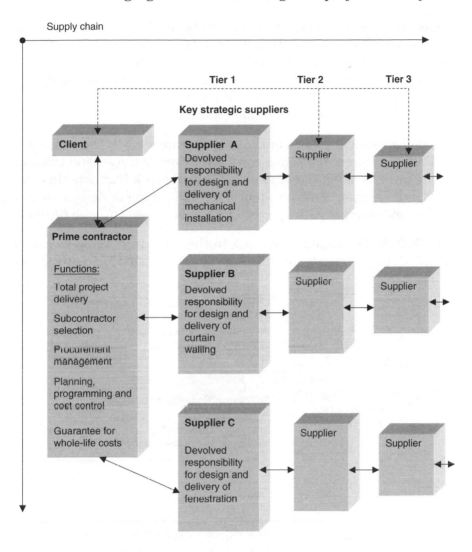

-----Step in agreement for client in areas of commercial benefit, e.g.-a super market chain's influence in the purchase of refrigeration equipment and plant.

Figure 3.3 Prime contracting.

stated, prime contracting requires there to be a single point of responsibility, this can be achieved through the prime contractor/subcontractor arrangements as the prime contractor takes full responsibility for the performance of its subcontractors and consultants. However, it is also necessary for the prime contractor to demonstrate an ability to bring together all the parties of the supply chain. This it is suggested could be achieved either by entering into partnering style alliances or alternatively a series of non-binding partnering protocols as discussed previously.

Several models of prime contracting procurement are being developed. Figure 3.4 illustrates the path used by the Ministry of Defence. The approach is characterised by:

- The early involvement of the prime contractor in the process.
- The selection of the prime contractor during the second stage of the process along the lines of a Public Private Partnership deal.
- The devolution of responsibility for the complete design, execution and delivery of the project with guaranteed through life cost.

The MoD in its prime contracting initiative is to develop so-called 'clusters' or elements of the supply chain that constitute an

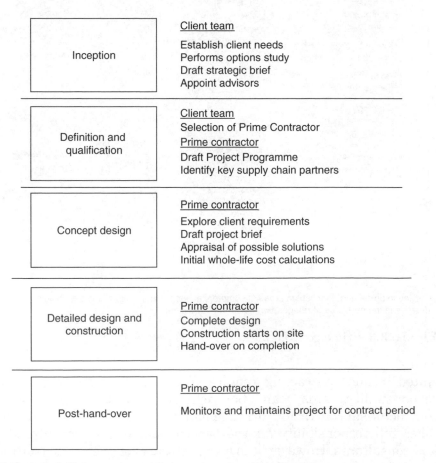

Figure 3.4 Prime contracting procurement process.

integrated team who would work together on a particular part of the works, for example:

- Groundworks
- Lift installation
- Roofing, etc.

These clusters, or tiers, are built outside particular projects and there could be two or three supply chains capable of delivering an outcome for each cluster. A typical cluster for say, mechanical and electrical services could include the design engineers, the contractor and the principal component manufacturers. Crucially, for the success of this approach, clusters must have the confidence to proceed in the design and production of their element in the knowledge that clashes in design or product development with other clusters are being managed and avoided by the prime contractor. Without this assurance then this approach offers little more than the traditional supply chain management techniques where abortive and unco-ordinated work is unfortunately the norm. It should be noted that the legal structure of such clusters has yet to be formalised.

Another important step in creating the right environment for prime contracting is the approach towards liability for defects. In the traditional approach, each party makes their own arrangements for indemnity insurance, whereas in some prime contracting models, blanket professional indemnity insurance cover is arranged for the whole supply chain, thereby eliminating the blame culture that pervades the industry and reinforcing the sense of well-being for the client. This approach should also theoretically reduce the overall costs of insurance.

The prime contractor's responsibilities might include the following:

- Overall planning, programming and progressing of the work
- Overall management of the work, including risk management
- Design co-ordination, configuration control and overall system engineering and testing
- The pricing, placing and administration of suitable subcontractors
- Systems integration and delivering the overall requirements

The role of the quantity surveyor in prime contracting to date has tended to be as works advisor to the client, a role which includes examination of the information produced by the prime contractor, a role it must be said resented by prime contractors, and as cost consultant to a prime contractor without the necessary in-house disciplines. However, there is nothing to prevent a quantity surveying consultancy from taking the prime contractor role, provided the particular requirements of the client match the outputs and that added value can be demonstrated.

Prime contracting in the public sector has a three/four stage process. A four-stage procurement process follows:

1. *Invitations for expressions of interest.* The normal advertising process with the OJEU (not required for the private sector).
2. *Pre-qualification questionnaire.* At this point, the interested parties are given sufficient information to understand the extent and scope of the services required and of the constraints under which they may be required to operate. Any potential prime contractor must be able to demonstrate that within their organisation there exists a well-established and proven supply chain management structure. It should be noted that after this stage bidding costs do start to increase for contractors and so five bidders will be selected to proceed to the next stage and submit bids.
3. *Invitations to tender.* The selected contractors are invited to submit full tenders and will include: concept drawings, stage 3 RIBA plan of work, value analysis of client's requirements, risk analysis and whole-life cost figures developed in conjunction with the supply chain partners.
4. *Preferred bidder and contract award.* After submission of the bids a preferred bidder as well as a reserve are selected.

The key commercial issue surrounding prime contracting is setting up long-term relationships based on improving the value of what the supply chain delivers, improving quality and reducing underlying costs through taking out waste and inefficiency. It is claimed by the MoD that the products and services provided by the companies in the supply chain typically account for 90 per cent of the total cost of a construction project. The performance of the whole supply chain impacts on the way in which the completed building meets the client's expectations. By establishing long-term relationships with supply chain members, it is believed that the performance of built assets will be improved through:

- the establishment of improved and more collaborative ways of working together to optimise the construction process
- exploitation of the latest innovations and expertise.

Sustainability/Whole-life costs

A recent report by the National Audit Office Improving Public Services Through Better Construction, concluded that £2.6 billion per annum is still wasted through a variety of reasons, including lack of consideration of whole-life cost and sustainability or green issues.

One of the major obstacles to the introduction of more sustainable design and construction solutions is that the belief to do so will involve additional costs – typically 10 per cent on capital costs. However a 2005 report by BRE Trust and Cyril Sweett entitled *Putting a Price on Sustainability*, appears to demonstrate that this need not necessarily be the case. In fact the report points out that significant improvements in the sustainability performance of buildings need not be expensive, although it is still true that in order to attain high EcoHomepoint or BREEAM ratings (see below for definitions) there is the need to incur significant up front investment. Nevertheless, the general uncertainty over the cost impact to an entire development's profitability could deter risk adverse funders from backing a green project.

Various attempts have been made to define the term 'sustainable or green construction'. In reality it would appear to mean different things to different people in different parts of the world depending on local circumstances. Consequently, there may never be a consensus view on its exact meaning; however, one way of looking at sustainability is:

> The ways in which built assets are procured and erected, used and operated, maintained and repaired, modernized and rehabilitated and reused or demolished and recycled constitutes the complete life cycle of sustainable construction activities.

In 2005, the RICS announced that it was establishing a new commission with a mission to 'Ensure that sustainability becomes and remains a priority issue throughout the profession and RICS'. In general, a sustainable building reduces the impact on the environmental and social systems that surround it, as compared to conventional buildings. Green buildings use less water and energy, as well as fewer raw materials and other resources.

Why should sustainability concern the quantity surveyor? Here are some statistics:

- The construction sector is responsible for one-sixth of the total freshwater withdrawals and taking into account demolition, generates 30 per cent of waste in OECD countries. In addition, around 40 per cent of total energy consumption and greenhouse gas emissions are directly attributable to constructing and operating buildings according to Energy Action. Measured by weight, construction and demolition activities also produce Europe's largest waste stream, between 40 per cent and 50 per cent, most of which is recyclable
- Contractors, particularly those involved with public private partnerships are recognising the importance of sustainability issues and the early consideration of whole-life costs
- Construction clients increasingly are realising the marketing potential of green issues and
- Thirty per cent of newly built or renovated buildings suffer from sick building syndrome.

The process of getting the minimum whole-life cost and environmental impact is so complex, being a three-dimensional problem as

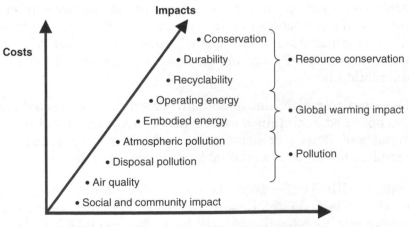

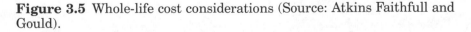

Figure 3.5 Whole-life cost considerations (Source: Atkins Faithfull and Gould).

indicated in Figure 3.5. Each design option will have associated impacts and costs, and trade-offs have to be made between apparently unrelated entities, e.g. what if the budget demands a choice between recycled bricks or passive ventilation.

The measures adopted to assess sustainability performance, encourage developers and design teams to consider the following issues at the earliest possible opportunity:

- EcoHomes points
- BREEAM (Building Research Establishment Assessment Method) rating.

EcoHomes points

EcoHomes assesses the green performance of houses over a number of criteria:

- Reducing CO_2 emissions from transport and operational energy
- Reducing main water consumption
- Reducing the impact of materials use
- Reducing pollutants harmful to the atmosphere and
- Improving the indoor environment.

BREEAM (Building Research Establishment Assessment Method)

BREEAM has been developed to assess the environmental performance of both new and existing buildings. BREEAM assesses the performance of buildings in the following areas:

- Management: overall management policy, commissioning and procedural issues
- Energy use
- Health and well-being
- Pollution
- Transport
- Land use
- Ecology
- Materials and
- Water, consumption and efficiency.

In addition, unlike EcoHome points, BREEAM covers a range of building types, such as:

- Offices
- Industrial units
- Retail units
- Schools and
- Other building types, such as leisure centres, can be assessed on ad hoc basis.

In the case of an office development, the assessment would take place at the following stages:

- Design and procurement
- Management and operation
- Post-construction reviews and
- Building performance assessments.

BREEAM measures the environmental performance of buildings by awarding credits for achieving a range of environmental standards and levels of performance. Each credit being weighted according to its importance and the resulting points are added up to give a total BREEAM score and rating.

Putting a price on sustainability

It is commonly assumed that consideration of sustainable issues will rack up the costs of a building, but this may not necessarily be the case. One of the principal barriers to the wider adoption of more sustainable design and construction solutions is the perception that these incur additional unwanted costs. Location and site conditions have a major impact on the assessment and of course these factors may be outside the design team's influence.

Popular perception is that there is a lack of customer demand for sustainability to be considered during design and procurement stages; however, consider the following reasons to be green:

- The Housing Corporation requires an EcoHomes 'Good' rating for any scheme they fund
- English Partnerships requires developers to achieve a minimum BREEAM or EcoHomes very good rating

- Public sector contractors must achieve BREEAM excellence for all new buildings
- Many high-profile private developers and landowners are seeking the same or higher standards of sustainability performance from their partners
- Investors are becoming increasingly interested in sustainability and are encouraging property industry partners to do the same.
- The imminent implementation of the Energy Performance of Buildings Directive (discussed later).

Some of the benefits of having a BREEAM rating are claimed to be:

- Demonstrating compliance with environmental requirements
- Marketing: as a selling point to tenants and customers
- Financial: to achieve higher and increased building efficiency.

A BREEAM rating assessment comes at a price and according the BRE website the fee scale for BREEAM assessors to carry out an assessment at each of the above stages could be as high as £10 000 per stage.

Value from green development

Comparatively few green buildings have been completed and of those a high percentage have been in the public sector. However, proponents of green development maintain that sustainable buildings can:

- Significantly reduced whole-life costs and ensure more rapid pay back compared to conventional buildings from lower operation and maintenance costs, thereby generating a higher return on investment
- Secure tenants more quickly
- Command higher rents or prices
- Enjoy lower tenant turnover
- Attract grants, subsidies, tax breaks and other inducements
- Improve business productivity for occupants.

However, the barriers to green development are at present substantial and include:

- Lack of a clear project goal – i.e. targets
- Lack of experience

- Lack of commitment
- Complicated rating process.

A raft of new legislation is now ensuring that the consideration of sustainability for new projects is not merely an option. In January 2006, the European Commission's Energy Performance Directive comes into effect and shortly after that in April 2006 the Office of the Deputy Prime Minister plans to launch a Code for Sustainable Buildings. If this were not enough, the long awaited revision to Part L of the Building Regulations also comes into force in April 2006. The Building Regulations are considered to be one of the most far reaching pieces of legislation ever to hit the construction industry and will force cuts in the carbon emissions from buildings by one million tonnes a year. Design teams will have to obtain energy ratings before and after construction and assessment will be based on the Government's Standard Assessment Procedure for Energy Rating 2005 (SAP 2005).

Whole-life costs

First introduced to the UK construction industry over four decades ago by Dr. P. A. Stone as Costs-in-Use, it is only recently, with the widespread adoption of public private partnerships as the preferred method of procurement by the majority of public sector agencies, that the construction industry has started to see some merit in whole-life costing. In addition building owners with long-term interests in property are starting to demand evidence of the future costs of ownership. For example, PFI prison projects are commonly awarded to a consortium on the basis of design, build, finance and operate (DBFO), and contain the provision that, at the end of the concession period, typically 35 years, the facility is handed back to the HM Prisons in a well-maintained and serviceable condition. This is, of course, in addition to the operational and maintenance costs that will have been borne by the consortia over the contract period. Therefore, for PFI consortia, given the obligations touched on above, it is clearly in their interest to give rigorous attention to costs incurred during the proposed assets life cycle in order to minimise operational risk. Although Stone's work was well received in academic circles, where today extensive research still continues in this field, there has been and continues to be, a good deal of apathy in the UK construction industry to the wider consideration of whole-life costs.

It has been estimated that the relationship between capital costs: running costs and business costs in owning a typical office block over a 30-year period are as shown in Table 3.3.

Although recently some doubt has been expressed as to the accuracy of the Royal Academy ratio and that in reality it could be closer to 1:2:200, the fact remains that whole-life costs are still a considerable factor in the cost of built asset ownership.

The business operating costs in the above equation (200) include the salaries paid to staff, etc. Clearly in the long term this aspect is worthy of close attention by design teams and cost advisors. One of the reasons for this lack of interest, particularly in the private sector and the developers' market, is that during the 1980s financial institutions became less enamoured with property as an investment and turned their attention to the stock market. This move led to the emergence of the developer/trader. Often an individual, rather than a company, who proposed debt finance rather than investment financed development schemes. Whereas previously, development schemes had usually been prelet and the investor may even have been the end user, the developer trader had as many projects in the pipeline as they could obtain finance for. The result was an almost complete disregard for whole-life costs as pressure was put on the designers to pare capital costs at the expense of ultimate performance as building performance is poorly reflected in rents and value. Fortunately, these sorts of deals have all but disappeared, with a return to the practice of preletting and a very different attitude to whole-life costs. If a developer trader was developing a building to sell on, they would have little concern with the running costs, etc.; however, in order to prelet a building, tenants must be certain that, particularly if they are entering a lease with a full repair and maintenance provision, there are no 'black holes', in the form of large repair bills, waiting to devour large sums of money at the end of the lease. In the present market therefore sustainability is as important to the developer as the owner/occupier. A building will have a better chance of attracting better quality tenants, throughout its life, if it

Table 3.3 Relationship between capital, running and business costs

Construction (capital) cost	1
Maintenance and operating costs	5
Business operating costs	200

Source: The long-term costs of owning and using buildings – Royal Academy of Engineering and Stanhope (1998).

has been designed using performance requirements across all asset levels, from Facility (building), through System (heating and cooling system), to Component (air handling unit), and even Sub–Component (fans or pumps).

In and around major cities today, it is clear that buildings that attracted good tenants and high rents in the 1980s and early 1990s are now tending to only attract secondary or tertiary covenants, in multiple occupancies, leading to lower rents and valuations. This is an example of how long-term funders are seeing their 25–35-year investments substantially underperforming in mid-life, thus driving the need for better whole-life procured buildings.

Whole-life cost procurement includes the consideration of the following factors:

- *Initial* or procurement costs, including design, construction or installation, purchase or leasing, fees and charges
- *Future* cost of operation, maintenance and repairs, including: management costs such as cleaning, energy costs, etc.
- *Future* replacement costs including loss of revenue due to non-availability
- *Future* alteration and adaptation costs including loss of revenue due to non-availability
- *Future* demolition/recycling costs.

Whole-life appraisers may include whatever they deem to be appropriate – provided they observe consistency in any cross comparisons. The timing of the future costs associated with various alternatives must be decided and then using a number of techniques described below assess their impact. Classically, whole-life cost procurement is used to determine whether the choice of, say, a component with a higher initial cost than other like-for-like alternatives is justified by being offset by reduction of the future costs as listed above. This situation may occur in new build or refurbishment projects. In addition, whole-life cost procurement can be used to analyse whether, in the case of an existing building, a proposed change is cost effective when compared against the 'do nothing' alternative.

There are three principal methods of evaluating whole-life costs:

- Simple aggregation
- Net present value
- Annual equivalent approach.

Simple aggregation

The basis of whole-life costs is that components or forms of construction that have high initial costs will, over the expected life span, prove to be cheaper and hence better value than cheaper alternatives. This method of appraisal involves adding together the costs, without discounting, of initial capital costs, operation and maintenance costs. This approach has a place in the marketing brochure and it helps to illustrate the importance of considering all the costs associated with a particular element but has little value in cost forecasting. A similarly simplistic approach is to evaluate a component on the time required to pay back the investment in a better quality product. For example, a number of energy saving devices are available for lift installations, a choice is made on the basis of which over the life cycle of the lift, say 5 or 10 years, will pay back the investment the most quickly. This last approach does have some merit, particularly in situations where the life cycle of the component is relatively short and the advances in technology and hence the introduction of a new and more efficient product likely.

Net present value approach

The technique of discounting allows the current prices of materials to be adjusted to take account of the value of money of the life cycle of the product. Discounting is required to adjust the value of costs, or indeed, benefits which occur in different time periods so that they can be assessed at a single point in time. This technique is widely used in the public and the private sectors as well as sectors, other than construction. The choice of the discount rate is critical and can be problematic as it can alter the outcome of calculation substantially. However, when faced with this problem, the two golden rules that apply are that in the public sector follow the recommendations of the Green Book or Appraisal and Evaluation in Central Government, currently recommending a rate of 3.5 per cent. In the private sector, the rule is to select a rate that reflects the real return currently being achieved on investments. To help in understanding the discount rate, it can be considered almost as the rate of return required by the investor which includes costs, risks and lost opportunities.

The mathematical expression used to calculate discounted present values are set out below:

$$\text{Present value (PV)} = \frac{1}{(1+i)^n}$$

where i is rate of interest expected or discount rate and n is the number of years.

This present value multiplier/factor is used to evaluate the present value of sums, such as replacement costs that are anticipated or planned at, say, 10 or 15-year intervals.

For example, consider the value of a payment of £150 that is promised to be made in 5 years time:

Assuming a discount rate of 3.5 per cent, £150 in 5 years time would have a present worth or value of £126.30:

$$\frac{1}{(1+i)^n}$$

$$£150 \times \frac{1}{(1.035)^n} = £150 \times 0.8420 = £126.30$$

or in other words, if £126.30 were to be invested today at 3.5 per cent this sum would be worth £150.00 in 5 years time, ignoring the effects of taxation.

Calculating the present value of the differences between streams of costs and benefits provides the net present value (NPV) of an option and this is used as the basis of comparison as outlined in the next section.

Annual equivalent approach

This approach is closely aligned to the theory of opportunity costs, i.e. the amount of interest lost by choosing option A or B as opposed to investing the sum at a given rate percentage, is used as a basis for comparison between alternatives. This approach can also include the provision of a sinking fund in the calculation in order that the costs of replacement are also taken into account.

In using the annual equivalent approach the following equation applies:

Present value of £1 per annum (sometimes referred to by actuaries as the annuity that £1 will purchase)

This multiplier/factor is used to evaluate the present value of sums, such as running and maintenance costs that are paid on a regular annual basis

$$\text{Present value of £1 per annum} = \frac{(1+i)^n - 1}{i(1+i)^n}$$

where i is the rate of interest expected or discount rate and n is the number of years.

Previously calculated figures for both multipliers are readily available for use from publications such as Parry's *Valuation Tables*, etc.

Sinking funds should also be considered; a fund created for the future cost of dilapidations and renewals. Given that systems are going to wear out and/or need partial replacement it is thought to be prudent to 'save for the rainy day' by investing capital in a sinking fund to meet the cost of repairs, etc. The sinking fund allowance therefore becomes a further cost to be taken into account during the evaluation process. Whether this approach is adopted will depend on a number of features including corporate policy, interest rates, etc.

Whole-life costing is not an exact science, as, in addition to the difficulties inherent in future cost planning, there are larger issues at stake. It is not just a case of asking 'how much will this building cost me for the next 50 years', rather it is more difficult to know whether a particular building will be required in 50 years time at all – especially as the current business horizon for many organisations is much closer to 3 years. Also, whole life costing requires a different way of thinking about cash, assets and cash flow. The traditional capital cost focus has to be altered, and costs be thought of in terms of capital and revenue costs coming from the same 'pot'. Many organisations are simply not geared up for this adjustment. The common misconception that a whole-life costed project will always be a project with higher capital costs does not assist this state of affairs. As building services carries a high proportion of the capital cost of most construction projects, this is of particular importance. Just as capital and revenue costs are intrinsically linked so are all the variables in the financial assessment process. Concentrate on one to the detriment of the others and you are likely to fail.

Perhaps, the most crucial reason is the difficulty in obtaining the appropriate level of information and data.

The lack of available data to make the calculations reliable. Clift & Bourke 1999 found that despite substantial amounts of research into the development of database structures to take account of performance and WLC there remains significant absence of standardisation across the construction industry in terms of scope and data available. Ashworth also points out that the forecasting of building life expectancies is a fundamental prerequisite for whole-life cost calculations, an operation that is fraught with problems. While to some extent building life relies on the lives of the individual building components, this may be less critical than at first imagined, since the major structural elements, such as the substructure and the frame, usually have a life far beyond those of the replaceable elements. Clients and users will have theoretical norms of total life spans but these have often proved to be widely inaccurate in the past. Building Maintenance Information of the Royal Institution of Chartered Surveyors was established in the 1970s. BMI have developed a standard form of property occupancy cost analysis, which it is claimed allows comparisons between the cost of achieving various defined functions or maintaining defined elements. The BMI define an element for occupancy cost as: expenditure on an item which fulfils a specific function irrespective of the use of the form of the building. The system is dependent on practitioners submitting relevant data for the benefit of others. The increased complexity of construction means that it is far more difficult to predict the whole-life cost of built assets. Moreover if the malfunction of components results in decreased yield or underperformance of the building then this is of concern to the end user/owner. There is no comprehensive risk analysis of building components available for practitioners, only a wide range of predictions of estimated life spans and notes on preventive maintenance – this is too simplistic. There is a need for costs to be tied to risk including the consequences of component failure. After all the performance of a material or component can be affected by such diverse factors as:

- Quality of initial workmanship when installed on site and subsequent maintenance.
- Maintenance regime/wear and tear. Buildings that are allowed to fall into disrepair prior to any routine maintenance being carried out will have a different life cycle profile to buildings that are regularly maintained from the outset.

- Intelligence of the design and the suitability of the material/ component for its usage. There is no guarantee that the selection of so-called high quality materials will result in low life cycle costs.

Other commonly voiced criticisms of whole-life costs (WLC) are:

- Expenditure on running costs is 100 per cent allowable revenue expense against liability for tax and as such is very valuable. There is also a lack of taxation incentive, in the form of tax breaks, etc., for owners to install energy efficient systems.
- In the short term and taking into account the effects of discounting, the impact on future expenditure is much less significant in the development appraisal.

Another difficulty is the need to be able to forecast, a long way ahead in time, many factors such as life cycles, future operating and maintenance costs, and discount and inflation rates. WLC, by definition, deals with the future and the future is unknown. Increasingly obsolescence is being taken into account during procurement – a factor that it is impossible to control since it is influenced by such things as fashion, technological advances and innovation. An increasing challenge is to procure built assets with the flexibility to cope with changes. Thus, the treatment of uncertainty in information and data is crucial as uncertainty is endemic to WLC (Flanagan *et al.*, 1989; Bull, 1993). Another major difficulty is that the WLC technique is expensive in terms of the time required. This difficulty becomes even clearer when it is required to undertake a WLC exercise within an integrated real-time environment at the design stage of projects.

Nevertheless, changes in the nature of development mean that other factors have emerged to convince the industry that whole-life costs are important.

Whole-life cost procurement – critical success factors:

- Effective risk assessment – what if this alternative form of construction is used?
- Timing – begin to assess WLC as early as possible in the procurement process
- Disposal strategy – is the asset to be owner occupied, sold or let?
- Opportunity cost – downtime
- Maintenance strategy/frequency – does one exist?

- Suitability – matching a client's corporate or individual strategy to procurement.

In 2001, the Whole-Life Cost Forum was launched as a source of reference on whole-life cost data and can be accessed at www.wlcf.org.uk.

Conclusion

Although sustainability/green development may still for many seem to lie in the field 'crankiness', there are undoubtedly strong moves to increase the awareness of the impact that pollution and high energy consumption are having on the world and as, for example, concerns for climate change increase there will be increased pressure on quantity surveyors to provide data on the costs of going green.

Bibliography

Bennett, J. and Jayes, S. (1995). *Trusting the Team – The Best Practice Guide to Partnering in Construction*. Centre for Strategic Studies in Construction, The University of Reading.

Bennett, J. and Jayes, S. (1997). *The Seven Pillars of Partnering – A Guide to Second Generation Partnering*. Centre for Strategic Studies in Construction, The University of Reading.

Blois, K. J. (1999). Trust in business to business relationships: an evaluation of its status. *Journal of Management Studies*, 36, 197–215.

BRETrust, Cyril Sweett (2005). *Putting a Price on Sustainability*, BRE Bookshop.

Building (2005). Datafile, *Building* 6 May, pp. 85–89.

Bull, J.W. (1993). The way ahead for life cycle costing in the construction industry. In: Bull J.W. (ed.), *Life Cycle Costing for Construction*. Blackie, Glasgow.

Construction Industry Training Board, Working Group 12 (1997). *Partnering in the Team*, Thomas Telford Publishing.

Davies R. Dr (ed.) (2005). Green value report, Royal Institution of Chartered Surveyors.

DETR (1998). *Rethinking Construction*. Report of the Construction Task Force to the Deputy Prime Minister, John Prescott, on the scope for improving the quality and efficiency of UK construction, Department of the Environment, Transport and the Regions.

Fisher, N. and Green, S. (2001). Partnering and the UK construction industry, the first ten years – a review of the literature. In: *Modernising Construction*, Appendix 4. HMSO.

Flanagan, R., Norman, G., Meadows, J. and Robinson, G. (1989). *Life Cycle Costing – Theory and Practice*. BSP Professional Books.

Heywood, A. (2005). Green light for sustainability, *RICS Business* 16–17.

HM Treasury (1999). Procurement Guidance No. 4: Teamworking, Partnering and Incentives. Procurement Group.

Howell, G. *et al.* (2002). *Beyond Partnering: Towards a New Approach to Project Management.*Lean Construction Institute, Stanford University.

NAO (2005). *Improving Public Services Through Better Construction*, National Audit Office.

National Economic Development Office (1991). *Partnering: Contracting Without Conflict*. HMSO.

Ng, S. T. *et al.* (2002). Problematic issues associated with project partnering – the contractor perspective, *International Journal of Project Management*, 20, 437–449.

Richardson, S. (2005). Public sector wastes £2.6 bn yearly, *Building* 18 March, pp. 16.

Wood, G. (2005). Partnering practice in the relationship between clients and main contractors, *RICS* Research.

4

Procurement – doing deals

Introduction

This chapter examines the impact of public private partnerships (PPPs) and in particular the private finance initiative (PFI); a procurement strategy described by the Royal Institution of Chartered Surveyors, as one of the most important influences on the future of the quantity surveyor during the next decade; a sentiment echoed in 2005 in the RICS report Quantifying Quality. Successful PPPs depend upon many of the skills offered by quantity surveyors, for example, risk management, procurement advice, whole-life costs advice (previously discussed in Chapter 3), etc. The origins, philosophy and motives behind the introduction of public private partnerships are discussed as well as procedural and contractual aspects of PPPs, the PFI and the Public Private Partnership Programme (4Ps). The chapter will explore the claims and counter claims put forward by all sides for PPPs and will conclude by discussing the opportunities that PPP/PFI presents for quantity surveyors and the trend in the global adoption of PPPs.

Public private partnerships

Background and definition

In its widest sense, a public private partnership can be defined as 'a long term relationship between the public and private sectors that has the purpose of producing public services or infrastructure' (Zitron, 2004). One of the many PPP models (see Figure 4.1), is the private finance initiative (PFI); a term used to describe the procurement processes by which public sector clients contract for capital intensive services from the private sector. Private sector involvement in the delivery of public services in the UK has developed into a very emotive topic, with an unfortunate tendency to

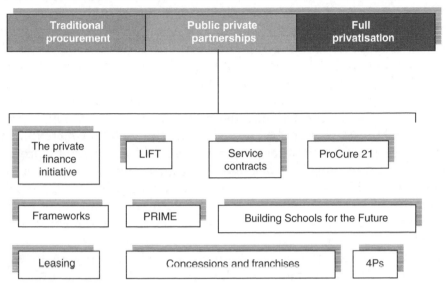

Figure 4.1 PPP procurement models.

generate more heat than light. For many, the confusion and mis-
conceptions surrounding PPP/PFI begins with the definition of
these two terms. As discussed later in this chapter, the term PFI
was launched in the early 1990s and then several years later the
term PPP emerged and appeared to subsume the PFI. Public
private partnerships bring public and private sectors together in
long-term contracts. PPPs encompass voluntary agreements and
understandings, service level agreements, outsourcing and the PFI.

Perhaps the adoption of the generic term public private partnerships in 1997 also had something to do with creating a softer image
of public and private sectors working together, sharing the risk and
rewards, as well as an attempt to counter the public perception that
to date, the PFI has tended to be synonymous with the private sector raping public sector services, such as the NHS and education,
making huge profits for shareholders, whilst imposing onerous
working conditions on staff running the services.

A PPP project therefore, usually involves the delivery of a traditional public sector service and can encompass a wide range of
options, one of which is, the PFI. In turn, the PFI is one of several
similar approaches in a 'family' of procurement that includes:
ProCure 21, LIFT, Building Schools for the Future, etc. (see Figure
4.1). One of the key objectives of the PFI is to bring private sector

management expertise and the disciplines associated with private ownership and finance into the provision of public services. However, if the PFI is to deliver value for money to the public sector, the higher costs of private sector finance and the level of returns demanded by the private sector investors must be outweighed by lower whole-life costs and increased risk transfer.

The range of PPP models being used in the provision of asset-based services in the UK include, as has already been noted, PFI deals and these are described in some detail in the following text, but there are a number of other potential forms of PPP including profit-sharing arrangements and long-term contracting, or a project in which the public sector provides the finance for a project and the private sector designs, builds and operates the asset. Examples of PPPs are the Channel Tunnel Rail Link, made possible by a government-backed bond and the modernisation of London Underground, which involves a private consortium taking over responsibility for investment in and maintenance of the infrastructure assets under a 25–30-year contract, while the public sector still remains in control of the operation of services.

The development of PPPs

In April 2004, in its Green Paper, *On Public–Private Partnerships and Community Law on Public Contracts and Concessions,* the European Commission used the term 'phenomenon' to describe the spread of public private partnerships across Europe. As will be discussed later in the chapter, PPPs and in particular the PFI, are now a global procurement model in which the UK is a world leader in terms of experience and know-how. Jimmi Bradbury, International Director, Cyril Sweett explains the recent increased interest in quantity surveyors in America for example to the realisation that PPPs and the PFI has so much to offer a country like the USA, where budget deficits are so high and public sector capital so limited. A particular case in point and a suitable case for PFI treatment is the state of California; the world's fifth largest economy, but with massive budget deficits that makes it impossible, using conventional funding and procurement techniques, to replace crumbling infrastructure.

The origins of the UK Private Finance Initiative lie in the introduction, in 1981, of the Ryrie Rules, after Sir William Ryrie, a former Second Permanent Secretary to the Treasury. These rules were ostensibly a means of allowing private financing and were developed to try to minimise the impact of government funding restrictions on possibly profitable investment by the nationalised industries. The rules were constantly criticised for being restrictive and gave public bodies little incentive to seek private finance alternatives and consequently there was almost no use of private finance in infrastructure projects until the construction of the Channel Tunnel in 1987. In an attempt to stimulate increased awareness a Green Paper, *New Roads by New Means* was published in 1989 by the Department of Transport on privately financed roads. The Ryrie Rules were partially phased out in 1989 and finally abandoned in 1992 with the launch of the PFI.

The private finance initiative is the name given to the policies announced by the Chancellor of the Exchequer, Norman Lamont in the autumn statement of 1992. The autumn statements of 1993 and 1994 by Chancellor Kenneth Clarke were used to reshape the design and nature of the initiative. The intention was to bring the private sector into the provision of services and infrastructure, which formerly had been regarded as primarily a public sector concern. For many political spectators, PFI was a natural progression for the Thatcher Government that had so vigorously pursued a policy of privatisation during the 1980s. In the UK from the end of the Second World War to 1980 autonomous agencies with political supervision, but not control, had been responsible for the delivery of electricity, gas, water, telecom, etc. These utilities were publicly owned assets and in theory any dividends went to the government. In practice, many of the industries were failing and consequently heavily subsidised by the Treasury and taxpayer in order to maintain employment. Privatisation, it was claimed, would introduce the management skills of the private sector to industries with an ethos of jobs for life and low accountability. In addition, privatisation would open up monopoly markets to other players with consequent efficiency gains and the bonus of added value for the customer. The process was achieved by valuing the publicly owned assets and then selling them off to the public by means of a share issue open to everyone. The privatisation programme proved to be extremely popular with both public as well as politicians and share issues were heavily oversubscribed – the term 'Stakeholder Britain' was coined to described the new phenomenon of widespread share ownership.

However, behind the queues of people waiting at stockbrokers' offices to deliver last minute applications for allocation of shares, there was the less publicised political agenda that continues to make privatisation so appealing to politicians and has led to its spread around the globe like a rash. At the time of the privatisation programme the public sector borrowing requirement (PSBR), latterly replaced by the so-called 'golden rules', was the benchmark of a governments' ability to control public expenditure. It represented the amount of money needed to be borrowed by government to fund capital projects; the lower the PSBR, the more prudent the government, a point not missed by the electorate whose taxes in the main funded the capital works programmes. If a government needs to raise revenue therefore, privatisation was and still is a very easy way of doing it, as the whole of the proceeds of the sale of the once publicly owned asset can be set against government debts, reducing budget deficits in the short to medium term and thereby lowering the borrowing and taxation levels. It has been estimated that the British Government raised £60 billion by selling previously public owned assets, or what some critics of the programme regarded as the UK's family silver during the 1980s and early 1990s. Opponents of privatisation also claimed that the money raised could have been substantially higher as nearly all the public utilities were undervalued by as much as 50 per cent in the government's dash for cash. Not surprisingly, therefore, the PFI has been seen by some as a means of back door privatisation of public services and trade unions, in particular UNISON, have voiced their concerns over the adoption of the PFI. However, as far as government is concerned there is a clear distinction between the sale of existing public assets, which they see as privatisation, and the PFI, which they do not.

It was against this backdrop therefore that in 1992 the PFI was launched and almost immediately hit the rocks. The trouble came from two sides; first, the way by which civil servants had traditionally procured construction works and services, left them without the experience, flexibility or negotiation skills to 'do deals', a factor that was to prove such an important ingredient for advancement of the PFI. In addition there was still a large divide and inherent suspicion between the public and private sectors and very little guidance from government as to how this divide could be crossed. There is also little doubt that there was a faction within the public sector that would have likely to see this public/private separation maintained. Hence, in 1993 a government body, the Private Finance Panel was created to encourage the use of the PFI and further attempts were made in 1994 to ensure engagement of the public sector with PFI when the then Chancellor, Kenneth Clarke, made it plain that Treasury approval for capital projects would not be given unless they had been tested against the private finance model.

The second major problem in trying to get the PFI off the ground related to the way in which a whole range of projects in the early days of the initiative were earmarked by over zealous civil servants as potential PFI projects, when they were quite obviously not. The outcome of this was that consortia could spend many months or even years locked into discussions over schemes with little chance of success, because the package under negotiation failed to produce sufficient guaranteed income to pay off the consortia's debt due to onerous contract conditions and inequitable risk transfer stipulations by the public sector. This practice earned PFI the reputation of incurring huge procurement costs for consortia and contractors before it became apparent that the business case for the project would not hold water. The procurement costs were non-recoverable by the parties concerned and before long PFI earned the reputation of being procurement of the last resort, at least by the private sector. Figure 4.2 illustrates the cost differential between traditional tendering and PFI procurement.

A recent RICS report called for the government to reimburse unsuccessful PFI tenderers, but to date signs are that there will be no change from the current policy of letting the private sector bear the whole of the risk for bid costs. In the mid-1990s, the trade press ran several vigorous campaigns to revise and streamline the PFI among

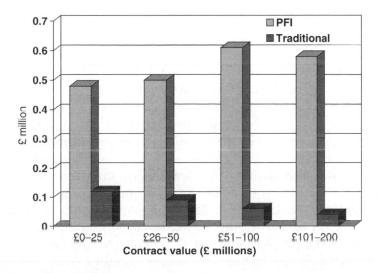

Figure 4.2 PFI bid costs.

reports that contractors had incurred costs of millions of pounds during abortive PFI negotiations. By October 1996, Bovis, Taylor Woodrow and other large contractors abandoned PFI bids, citing bureaucratic delays. The 1997 Labour Government was elected to power on a pledge to put partnership at the heart of modernising public services. Within a week of winning the election in May 1997, a Labour Government appointed Malcolm Bates to conduct a wide-ranging review of the PFI. The first Bates Report made 29 recommendations of which the dissolving of the Private Finance Panel in favour of a Private Finance Taskforce was key. The taskforce had two wings: an advisory wing and a projects wing. The advisory wing developed a standard learning package for the public sector and helped greatly to disseminate information about the PFI procurement process, which until then had seemed something of a black art. Following the publication of the second Bates Report in 1999 and its recommendation that deal-making skills could be strengthened and that all public sector staff engaged in PFI projects should undergo annual training, PricewaterhouseCoopers were tasked with producing a PFI competence framework. The object of the report and the framework was to identify the competences the public sector is likely to require in order to deliver successful PFI projects.

In 1999, Sir Peter Gershon was invited to review civil procurement in central government. The subsequent report highlighted a number of weaknesses in government procurement systems as follows:

- Organisation
- Process
- People and skills
- Measurement and
- Contribution of the central government.

Gershon's aim was to modernise procurement throughout government, provide a greater sense of direction in procurement and promote best practice in the public sector. Gershon's proposals for dealing with these deficiencies led to the creation of a central organisation entitled the Office of Government Commerce (OGC).

In June 2000, Partnerships UK (PUK) was established following the publication of the second Bates Report in 1999. Partnerships UK replaced the projects wing of the Treasury taskforce as a joint venture between the public and private sectors with the private sector holding the majority 51 per cent interest – it is itself a PPP!

The mission of PUK is to provide expertise to the public sector in order to provide better value for money for PPPs. Included in its remit is the sourcing and provision of finance or other forms of capital where these are not readily available from established financial markets, it makes a charge for its services. Most significantly PUK can be seen to mark a move towards greater centralisation in the management of PPP projects and the development of standard documents, including contracts, in direct contrast to the mid-1990s when each government department was encouraged to develop its own specialist expertise (Figure 4.3).

However, government was also anxious to spread the use of private investment into local authorities and in April 1996 the Local Authorities Association established the Public Private Partnership Programme or 4Ps in England and Wales. The 4Ps is a consultancy set up to help local authorities develop and deliver PFI schemes and other forms of public private partnership. The local authority services covered by the 4Ps are, for example, housing, transport, waste, sport and leisure, education, etc.

During the second and third terms of the Labour Government in the UK, a number of specialist PPP procurement routes have been devised in order to meet the needs of particular public sector agencies, as follows;

Partnerships for schools

Capital investment in schools in England is set to rise to over £5 billion in 2005/06. In 2004, the government issued its policy for the

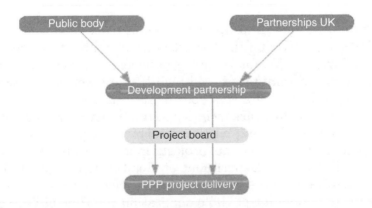

Figure 4.3 Development partnership agreement.

school building programme entitled, 'Building Schools for the Future' (BSF) which outlined the aims of the programme. Subsequently, PUK and the DfES established partnerships for schools to manage the delivery of the programme and developed the model now referred to as local education partnership (LEP). A LEP is a PPP between a local authority and a private sector partner, selected in open competition under EU public procurement rules referred to in Chapter 7.

The private sector partner (PSP) will own 80 per cent of the shares and be responsible for the management and delivery of a range of services. The PSP will be selected in open competition and may be a single company or a consortium of several companies specially formed to deliver the services required. The relationship will be long term with a view to developing the entire investment programme. The PSP will be selected based on its ability to provide partnering services and on the quality of its proposals. One of the biggest advantages with this approach to school building procurement is that the LEP will act as the single point of contact for the delivery of the design, construction, project management and maintenance. The LEP will oversee not only the delivery of PPP projects but also conventionally funded routes. The LEP enters into a 10-year strategic partnering agreement (see Chapter 2) and recovers its costs and earns returns through the contracts that it successfully delivers. The PSP will earn returns through management fees based on developing and procuring each project as well as dividends from risk capital invested to deliver PPP and/or conventional contracts.

NHS local improvement finance trust (LIFT)

Similar to LEPs, local improvement finance trusts involves Partnerships UK plc (PUK) and the Department of Health forming a joint venture, Partnerships for Health, to encourage investment in primary care and community based facilities and services. LIFT has been developed to meet a very specific need in the provision of primary and social healthcare facilities in inner city areas, i.e. GP surgeries, by means of a long-term partnering agreement. In order to participate in the programme, projects must be within areas designated as LIFT by the Department of Health. Although LIFT is at present confined to the health sector, other sectors are looking closely at the model for possible adaptation to other public service provision.

LIFT is based on an incremental strategic partnership and is fundamentally about engaging a partner to deliver a stream of accommodation and related services through a supply chain, established following a competitive EU compliant procurement exercise. Rather like the approach adopted by framework agreements, there should be no need to go through a procurement process again for a bidder to undertake these additional projects. Therefore, just as in the case of ProCure 21, see below, there should be considerable savings in terms of cost and time over the duration of the partnership arrangement.

LIFT, therefore, can be said to be a combination of existing procurement models as follows:

- The PFI which is a long-term partnership for the construction and maintenance of a built asset on the basis of payment on availability. Note that unlike PFI projects LIFT does not usually include soft facilities management, such as cleaning.
- Strategic partnering delivers a range of projects over a long period, usually 20–25 years. By their very nature, the value of these relatively small schemes is low, therefore several schemes can be batched together thereby resulting in procurement savings.
- LIFT enables strategic and capital investment planning between public and voluntary agencies involved in delivering public services.
- Joint venture companies.

In common with most partnering agreements or charters, the ethos is one of collaboration and conciliation and is for a term of approximately 20 years. LIFTs are expected to provide serviced accommodation suitable for use by, in the case of the NHS, health and social care professionals and practitioners to deliver services. The accommodation may be provided from new, refurbished or existing premises. A lease is entered into between the LIFT company and the occupants. The responsibility for the whole of the life cycle management of the asset feeds through to a payment mechanism that provides for no payment if accommodation is not available. Although there is a limited amount of public sector funding available, PFI credits and revenue support can be applied for from the 4Ps. As far as finance is concerned, there is no prescribed capital structure; however it should be available on a timescale that does not significantly delay the delivery programme. LIFT will build and refurbish primary healthcare premises, which it will then own. When complete the accommodation will be leased to GPs as well as chemists, opticians, dentists, etc. The attraction for the private sector partner is long-term income stream from rental payments and possibly secondary revenue streams.

Frameworks

Framework agreements are being increasingly used to procure goods and services in both the private and public sectors. Frameworks have been used for some years on supplies contracts; however, in respect of works and services contracts, the key problem, particularly in the public sector has been a lack of understanding as to how to use frameworks, whilst still complying with legislation, particularly the EU Directives and the need to include an 'economic test' as part of the process for selection and appointment to the framework. In the private sector, BAA were the first big player to use of framework agreements and covered everything from quantity surveyors to architects and small works contractors. The EU public procurement directives define a framework as

> An agreement between one or more contracting authorities and one or more economic operators, the purpose of which is to establish the terms governing contracts to be awarded during a given period, in particular with regard to price and, where appropriate, the quality envisaged.

Frameworks are discussed in more detail in Chapter 7, Global markets.

ProCure 21

NHS ProCure 21 has been constructed by NHS Estates around four strands to promote better capital procurement by:

- Establishing a partnering programme for the NHS by developing long-term framework agreements with the private sector that will deliver better value for money and a better service for patients

- Enabling the NHS to be recognised as a 'best client'
- Promoting high quality design
- Ensuring that performance is monitored and improved through benchmarking and performance management.

In common with most large public sector providers, the NHS has suffered from the usual problems of schemes being delivered late, over budget and with varied levels of quality combined with little consideration to whole-life costs. One of the main challenges to NHS capital procurement is the fragmentation of the NHS client base for specific healthcare schemes, as it comprises several hundreds of health trusts who all have responsibility for the delivery of schemes and each having differing levels of expertise and experience in capital procurement. The solution to these problems was to develop an approach to procurement known as NHS ProCure 21 as a radical departure from traditional NHS procurement methods and its cornerstone of the massive capital investment programme in the NHS in the period up to 2010. The principle underpinning the ProCure 21 programme is that of partnering with the private sector construction industry.

Although ProCure 21 ran into difficulties following the re organisation of NHS Estates in 2004 and received criticism from the framework contractors for lack of work, subsequently refuted in 2006 by an independent inquiry, it was commended in the 2004 National Audit Office report, Improving Public Services, through better construction as an example of best practice.

The Private finance initiative

The primary focus for PFI to date has been on services sold to the public sector. There are three types or PFI transactions currently in operation. The private finance initiative is the widest used, most controversial and best known form of PPP, currently accounting for approximately 80 per cent of all expenditure on PPPs in the UK construction sector. PFI deals have been used in some of the most complex and expensive PPP projects to date, such as the 872 bed New Royal Infirmary, Edinburgh (NRIE) and generally fall into three categories:

1. Classic PFI
2. Financially freestanding projects
3. Joint ventures.

Classic PFI

Typically, the private sector finances, builds and then operates over a 30–60 year period, a traditional public sector asset and in return receives a unitary payment based on performance and availability. In some cases, e.g. prisons, the private sector also provides staffing. Utilising one of the most popular private finance initiative models, Design, Build, Finance and Operate (DBFO) (Figure 4.4) Consort Healthcare, a private sector consortia comprising service group BICC, the Royal Bank of Scotland and Morrison Construction, designed and built the NRIE between 1998 and 2002, including arranging and providing the debt finance. Since its opening in 2002, Consort Healthcare maintains the non-clinical hospital services,

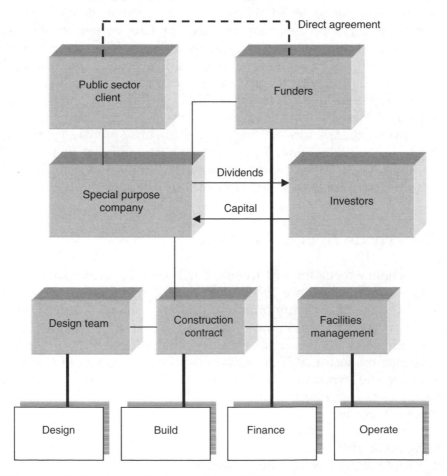

Figure 4.4 Contractual relationship – DBFO.

such as car parking, catering, cleaning, planned maintenance, etc. The public sector client, Lothian NHS Trust, retains responsibility for the clinicians and clinical services, including all medical staff. In return for providing and running the hospital building and all the ancillary services, Consort Healthcare receives a predetermined performance-based unitary payment for the duration of the PFI contract (30 years plus) providing, of course, that output and performance targets and standards are maintained and the NHS Trust continues to enjoy a state of the art hospital, including any commitments by Consort to refresh and update the technology and equipment during the contract period. The total value of the NRIE contract is £250 million over 32 years which includes not only the capital cost (£65 million) and finance costs, but also cost of the unitary charge. At the end of the contract period the hospital will be handed back, at no cost, to the NHS Trust in a good state of repair.

The classic PFI players

Special purpose vehicle/company

A special purpose vehicle is a consortium of interested parties brought together in order to bid for a PFI project. If successful, the consortium will be registered as a special purpose company, usually at the time of financial close. The special purpose vehicle/company or shell company as it is sometimes known, is a unique organisation constituted purely for a single PFI project. The company has a contractual link with the public sector sponsor or provider, the provider of finance, as well as the design, build and operating sectors of the project. In addition, the funder also usually has its own agreement with the sponsor which usually contains a step in clause, see Figure 4.4. This agreement is a safety net in the event that the special purpose company ceases trading or persistently fails to deliver services to the required contract standards. In the event of SPC failure then the whole project could be in jeopardy as the supply of finance to the organisation running the operating contract would stop. The step in clause permits finance to be channelled through the sponsors to ensure continuity in service delivery, which if interrupted, in the case of a hospital or school, would have wide-ranging consequences. In order to operate successfully, the special purpose company will have to call on the expertise of a wide range of consultants, including financial, legal, design, construction, facilities management, service delivery and employment law.

Public sector sponsor

The public sector client – experience has shown that one of the key roles within the PFI procurement process is the client team's project manager. Optimum progress is made when the project is managed by a person who has the time and authority to take decisions and negotiate with bidders, instead of having to keep referring back. The public sector client also usually relies heavily on input from consultants in the fields of procurement including EU procurement law (see Chapter 7), project planning, production of an output specification, evaluation of bids, drawing up contract documents and business cases, etc.

The funder

One of the defining characteristics of PFI procurement is that the project is financed from private sector sources instead of central public funds. Private finance will always be more expensive than funds from the UK Treasury, although this higher cost is said to be offset by the greater efficiencies introduced by private sector operators plus costs saved by the public sector when risks are transferred to the successful private consortia. In the final stages of the PFI procurement, the whole deal is put under the microscope in a process known as due diligence, which is often carried out by the funders. These checks often pose serious questions about risk and other aspects of the contract that purchasers and providers think they have already resolved. Banks rarely commit the time of their own staff until they have decided that they have enough of an interest in the scheme to make it worthwhile for their own legal experts to go through the document in detail. This final step can be a trial of nerves as this process can take several weeks or even months and can and frequently does involve previously agreed points being renegotiated to the satisfaction of the funders. Things can go wrong even at this late stage. In 2005, the collapse of the £1.1 billion PFI Paddington Health Campus Project left a legacy of £14 million of abortive costs, including £7.8 million of consultants' fees. The scheme, originally proposed in 1998, was scrapped after a key partner refused to approve the business case for the project although subsequently revived in 2006. Generally, as the PFI process has matured the attitude of the financial institutions has become less suspicious and generally more willing to fund projects. One of the results of this change in position is the reduction in the cost of funding and an increase in the sources. Funding options now include: on balance sheet (i.e. from existing corporate funds); equity and subordinated debt from consortium members; leasing or similar tax-driven instruments, the bond market (either wrapped or unwrapped); senior debt, normally provided by major banks on the same basis as normal project financing; and mezzanine or junior debt, normally supplied by specialist departments or subsidiaries of major banks and by independent specialists. It is now thought by many financial institutions that the differential between public and private sector finance is almost insignificant – a major shift in attitude that has led to early PFI projects being refinanced – see later reference in this chapter. Table 4.1 illustrates the trends in PFI financing during the past 10 years or so. In particular, the reduction in return on equity indicates a process that is maturing from the initial high-risk proposition to mainstream financing. Another aspect, illustrated in Table 4.1, is the high gearing of the PFI project; i.e. the high level of debt compared with equity. This has become higher recently as funders become more relaxed with the PFI process.

The design and construction team

The design and construct part of the process is usually the most straightforward and easily understood part of the procedure, with the majority of design teams and contractors leaving the project once the construction phase is completed and ready to start operating. One of the major criticisms of PFI projects has been their lack of architectural merit and design innovation. Some of the causes given are that the design period is too short, that there is too much commercial pressure and insufficient contact with user/client. A jewel in the PFI crown from the design perspective is the Anshen Dyer-designed Norfolk and Norwich Hospital, which opened in 2002 to critical acclaim (Figure 4.5).

Table 4.1 Trends in PFI financing

	Early PFI	Current market
Term	18 YEARS	25–30 YEARS
Debt/equity ratio (Gearing)	80/20	90/10
Annual debt cover	1.25	1.15
Return on equity	20–30%	12–15%

Figure 4.5 Norfolk and Norwich Hospital designed by Anshen Dyer (Photo Nick Cane).

The operator

The operator is the 'O' in DBFO and in the case of say, a school, this organisation will have complete devolved responsibility for the day-to-day operation of the facility, which may include such diverse services as the provision and maintenance of information technology provision, to lunches and playing field maintenance. In the case of ITCs, there is often a contractual obligation to ensure that the hardware and software are kept up to date with the latest versions of programmes and technologies. The operator clearly is the major player in ensuring project success and receives payment based on the quality and the reliability of the services.

Financially freestanding projects

The second PFI model is one in which the private sector supplier designs, builds, finances and then operates an asset, recovering costs entirely through direct charges on the private users of the asset, e.g. tolling, rather than payments from the public sector. Public sector involvement is limited to enabling the project to go ahead through assistance with planning, licensing and other statutory procedures. There is no government contribution or acceptance of risk beyond this point and any government customer for the specific service is charged at the full commercial rate. Examples of this kind of project include the Second Severn Bridge, the Dartford River Crossing and the Royal Armouries

Joint ventures

Joint ventures are projects to which both the public and private sectors contribute, but where the private sector has overall control. In many cases, the public sector contribution is made to secure wider social benefits, such as road decongestion resulting from an estuarial crossing. In other cases, government may benefit through obtaining services not available within the timescale required. The project as a whole must make economic sense and competing uses of the resources must be considered. The main requirements for joint venture projects are:

- Private sector partners in a joint venture should be chosen through competition
- Control of the joint venture should rest with the private sector
- The government's contribution should be clearly defined and limited. After taking this into account, costs will need to be recouped from users or customers
- The allocation of risk and reward will need to be clearly defined and agreed in advance, with private sector returns genuinely subject to risk.

The government's contribution can take a number of forms, such as concessionary loans, equity transfer of existing assets, ancillary or associated works, or some combination of these. If there is a government equity stake, it will not be a controlling one. The government may also contribute in terms of initial planning regulations or straight grants or subsidies.

Why the Private finance initiative?

There is nothing new with the involvement of the private sector in the delivery of public services. According to Zitron, in 1638 the Royal Navy considered, but subsequently rejected, the possibility of using a private consortium to store and provide meat for their men. In France, the Canal du Midi was completed in 1681 with the use of private finance. Similarly, there is nothing new in governments borrowing money from the private sector to finance the provision of public services and, in particular, the acquisition of capital assets. What is new, with a PFI deal for example, is that long-term (35–60 year) contracts, with the ownership and management of dedicated assets are left within the control of the private sector. The uniqueness of PPPs lies in the partnership of two sectors (public and private) which have during the last 60 years or so, in the UK at least, followed very different paths, with very different objectives. In broad terms, the benefit for the private sector includes the predictability of guaranteed long-term income streams and for the public sector, cost and time certainty in the delivery of a new or refurbished built asset that enables it to deliver a public policy outcome. In addition, given the unenviable track record of the UK construction industry, the public sector client does need to start paying for the facility or service until it is ready for use.

Until recently, government departments and agencies relied with varying degrees of success on traditional high-risk public procurement and funding models to provide the built assets necessary to deliver these public services. Typically, using traditional approaches government agencies draw up short and medium-term plans, prioritising needs and then attempt to arrange the finance, (from central government sources), design and construction of individual projects. The public sector agencies commonly utilise the services of the private sector for the design and construction, with the award of individual contracts being determined by a competitive bidding process based on lump sum contracts, often on the basis of drawings and a bill of quantities. Once the built asset has been completed, however, private sector participation, with a few exceptions noted below, ceases and the new facility is then operated and maintained by the public sector agency, together with other assets under its care. The difficulties with this fragmented approach to procurement have been three-fold:

1. Projects can only proceed once the public funding is in place and this can be problematic. Agencies have to bid annually, recently changed to 3-yearly, for funds from the Treasury and inevitably many projects fail to secure funding and do not go ahead. If funding is secured, design and procurement is usually on the basis of cheapest bottom line price rather than value for money, with little or no consideration given to long-term running, maintenance or decommissioning costs.
2. Once funding is approved, the project delivery is often unreliable both in terms of cost and time certainty, as previously discussed in Chapter 1.
3. The maintenance of built assets is also dependent on central government funding, which like the funding of capital projects is unpredictable. Often funds for capital building programmes have to be diverted to carry out essential maintenance or repair work.

The legacy of traditional procurement and funding strategies can best be illustrated by examining the National Health Service hospital building programme over the past 50 years. The NHS owns one of the largest and most complex property portfolios in Europe, with standing building stock worth £23 billion (disposal value) and £72 billion (replacement value), much of which is old and has been kept in service much longer than was originally envisaged, against a background of increasingly rapidly changing clinical techniques. Using traditional approaches to procurement and funding, out of 45 NHS major district hospital construction projects built between 1985 and 1996, the original tender price was exceeded by 10 per cent on 23 of these projects and by more than 20 per cent on 14 of them. In addition, 17 of the projects overran on time by over 10 per cent and in 10 of the projects by more than 20 per cent. More specifically, Guy's Hospital phase 3 was approved in principle in December 1986 at a cost of £35.5 million with a planned completion date of December 1993 – the reality was completion almost $3\frac{1}{2}$ years late and a final cost of £160 million, an increase of 450 per cent. The uncertainties of completion and final cost are one of the main difficulties of traditional procurement and makes even short-term planning very difficult. It is clear therefore that the traditional approach to public sector procurement, of keeping clear blue water between the public and private sectors was not working efficiently. During the preparation of the outline business case for the New Royal Infirmary in Edinburgh, one of the first PPP hospitals, independent construction companies advised the trust that if conventional procurement strategies were used, the project would take between 8 and 10 years, as compared with 4 years for a PPP design and build deal. Traditional procurement, has in many cases, not only delivered suboptimal performance in construction, but also in property maintenance. Prior to the redevelopment of West Middlesex University Hospital, using a PPP, the hospital trust estimated there was a backlog in maintenance of £22.8 million and in 1997 it was estimated that the bill for the backlog maintenance across the NHS as a whole was £3 billion.

In addition, traditional procurement models leave the public sector client vulnerable to high levels of risk which, it has been proved, it is ill equipped to manage. PFI procurement results in a large proportion of risk being transferred to the contractor or private sector.

Faced with patterns of inefficiency, the rationale for the introduction of alternative forms of procurement, with less risk for the public sector and that harness the expertise of the private sector, such as PPPs, seemed to need no explanation. Claims for the superior performance of non-state (private) institutions have been cited in relation to two key values. First, private sector organisations are viewed as being more efficient than public sector ones as a result of being more disciplined by market forces and competition. By comparison, public sector institutions are viewed as being excessively bureaucratic, controlled by administrative or professional interests and unresponsive to the pressures for efficiency which market-based organisations faced. Yet the opinions as to why PPPs are proving to be so popular with governments worldwide are truly diverse, as the following list demonstrates:

1. Government has an obligation to deliver public services, but government does not need to finance, build and maintain the infrastructure necessary to do this. Therefore at one level PPPs are a method of outsourcing the delivery of public services in which the government assumes the role of the purchaser of these services from the private sector supplier.

2. The European Commission takes a more pragmatic assessment, suggesting that PPPs are driven primarily by limitations in public funds to cover investment needs, but adding that

PPPs also boost efforts to increase the quality and roles for the private sector in PPP schemes:

- to provide additional capital
- to provide alternative management and implementation skills
- to provide value added to the consumer and the public at large
- to provide better identification of needs and optimal use of resources.

3. The Institute for Public Policy Research in its seminal report, Building Better Partnerships, came to the conclusion that the rationale for using PPPs instead of traditional approaches appeared to be confused, with muddled and contradictory statements by different government agencies.
4. PPPs allow projects to be funded throughout the economic cycle, which is attractive to governments taking a long-term view and anxious to keep public borrowing in check.

The debate goes on. However a constant theme concerning the rationale for the use of PPPs is that, in the medium term, they take some of the strain off public borrowing and add much needed certainty to cost and the delivery of public sector projects.

Compared with the traditional and often fragmented approaches to construction procurement, PPPs, depending on the model used, offer the advantages of synergies between traditionally diverse processes in the delivery and operation of built assets, for example:

- Synergy between the design and construction. This is not a new concept and buildability can also be achieved through other forms of procurement, such as design and build. Most PFI projects are able to deliver this well with designers working alongside the contractor
- Synergy between the construction phase and the operational phase. This is mainly to do with the suitability and reliability of the construction taking into account whole-life costs over the expected life of the project.

Not unnaturally, there is growing evidence that companies that can combine design, construction and hard facilities management in house, are increasingly successful in the PPP market, for example, the UK arm of the French giant Bouygues.

Accountability and public private partnerships

Traditionally, one of the major characteristics of traditional procurement is its transparency – an essential ethos of public sector finance. An event in December 2001 was to add fuel to the lobby which regarded PPP/PFI accounting practices with suspicion. The Enron Corporation was one of the world's largest energy, commodities and services company marketing electricity and natural gas, delivered energy and other physical commodities, as well as providing financial and risk management services to customers, worldwide. Based in Houston, Texas, Enron was formed in July 1985 by the merger of Houston Natural Gas and InterNorth of Omaha, Nebraska. Enron rapidly evolved from delivering energy to brokering futures as energy markets were deregulated. The company entered the European energy market in 1995. In 1999, Enron launched a plan to buy and sell access to high speed internet bandwidth and launched EnronOnline, a web-based commodity trading site. In 2000, Enron reported revenues of $101 billion and held stakes in 30 000 miles of gas pipeline and had access to a 15 000 mile fibreoptic network. On the face of it Enron was a success. In fact, it was known as one of the coolest organisations to work for with its executives leading expensive and lavish life styles and of course Enron's accounts were audited by one of the world's most respected firm of accountants, Arthur Anderson. However, disaster was just around the corner and in December 2001 following a collapse in its share price Enron, filed for Chapter 11 bankruptcy protection with an estimated $27 billion worth of what is known as off balance sheet debt. According to Iain McWhinney, President, Currie & Brown USA,

> There is definitely a legacy of Enron. Auditors are wanting to see greater cost visibility, as well as clear processes and procedures to manage it. This is where the QSs come into their own.

Many PFI deals are accounted for by the same off balance sheet processes used by Arthur Anderson when auditing Enron's accounts. Off balance sheet is an accounting procedure that for example, helps firms with significant long-term operating lease arrangements and unconsolidated affiliates, have debt-equity ratios that make them look financially healthier and more solvent than they actually are. Similarly, firms with resources at risk, because they have made certain guarantees that protect their customers or the creditors of unconsolidated affiliates or because they have left other contingent obligations off the balance sheet, may look like they have less debt than they actually have. By the same token, firms with large amounts of unfunded and unrecorded post-retirement obligations may look deceptively healthy.

The government publication, PFI: Meeting the Investment Challenge points out that 57 per cent of PFI projects by value are accounted for on balance sheet thereby leaving 43 per cent off balance sheet. In September 1998, the Accounting Standards Board (ASB) stated that the capital value of PFI transactions should appear on the government's balance sheet. However, negotiations between the ABS and the Treasury resulted in a revised version of the note, '*How to Account for PFI Transactions*', which allowed most PFI deals to be excluded from the government's borrowing figures on the grounds that they are 'operating leases'. In 1999 new guidance was issued which stated that property risks and service related risks (staffing costs) of PFI deals should be separated out. It should then be clear that the property-related risk has been transferred to the private sector and hence should not be on the government balance sheet. In 2001, the Office for National Statistics began to examine the ways in which PFI projects are treated for accounting purposes and by May 2005 was under pressure to make material changes in the way in which projects are classified. Any change from current *FRS 5 Accounting Standards Agency – Reporting the Substance of Transactions* processes to put more of the £43 billion spent of PFI project on balance sheet, would have the effect of raising public sector net debt to near the Treasury's self imposed limit of 40 per cent and making the PFI process far less attractive to public sector agencies and private consortia alike.

When launched in 1992 the mission of the PFI was stated by Chancellor Norman Lamont as 'allowing the private financing of capital projects'. In a speech to the annual PFI conference in October 1996, Chancellor Kenneth Clarke said, 'The injection of private capital investment and expertise that the PFI brings is a key factor in reconciling the need for sound public finances with acceptable levels of taxation and huge investment in infrastructure'. In 2001, Andrew Smith, the Chief Secretary to the Treasury wrote, 'It is central to the government's approach to use the PFI/PPPs over traditional forms of procurement only where they provide better value compared to traditional public provision. But better value for money means that we can deliver more essential services and to a higher standard than otherwise would be the case'. Between the delivery of these statements there lies almost a decade of confusion and exaggeration about the rational for PFI. Claim and counterclaim about its worth, as well as very real concerns about the ethics of allowing private organisations to build, operate and maintain the means of delivering healthcare and education for profit.

Whatever the confusion, the motive behind PFI remains the same: to re-align the public sector from being an owner of assets to a procurer of services. The government has a duty to provide and procure high quality public services in sectors such as healthcare, education, custodial services, etc. However, government does not have any need to own, maintain or operate the built assets that are essential to provide these services. Initial enthusiasm for the PFI was perhaps bordering on a 'something for nothing' mentality, that is to say instead of the government paying £65 million up front for a new prison, it was much better to get the private sector to provide, maintain and operate it. However, there truly is no such thing as a free lunch and it should not be overlooked that the PFI leaves a legacy in the payments that must be made under long-term contractual arrangements by successive governments. Table 4.2 illustrates the estimated payments that must be made by the Treasury under existing PFI contracts until year 2019 although contracts have been signed committing successive governments to make payments until 2031.

The current state of the PPP/PFI

Expenditure on PFI projects is in addition to traditionally centrally funded and procured public projects. In the NHS, for example, approximately 85 per cent of all projects are currently funded directly by the Treasury. By the end of 2005, approximately 680 PFI deals had been signed with a total value of £56 billion of which two major projects, the London Underground PPP and the

Table 4.2 Scale of PFI unitary payments

	£ million		*£ million*
2001–2002	2900	2010–2011	3500
2002–2003	3600	2011–2012	3600
2003–2004	3600	2012–2013	3600
2004–2005	3600	2013–2014	3400
2005–2006	3500	2014–2015	3400
2006–2007	3700	2015–2016	3100
2007–2008	3700	2016–2017	3200
2008–2009	3600	2017–2018	3200
2009–2010	3500	2018–2019	2700

Table 4.3 Investment in public sector projects

	2004/05	*2005/06*
	£ billion	*£ billion*
Public sector investment	33.0	42.0
Investment in PFI deals	2.8	3.2
Percentage of public expenditure	7.85	7.13

Channel Tunnel Rail Link, contributed a massive £32 billion. Of the traditional public service agencies, the Department of Health leads the way with £3600 million committed to deals in 2005/06 with further plans extending to 2008. In terms of overall public sector investment, the PFI has accounted for between 12 and 15 per cent of annual public sector capital investment although this is predicted to drop sharply in the short to medium term. Table 4.3 illustrates investment in public sector projects.

So, what is the current state of health of the PFI and why is it used? When it was first launched in 1992 the principle rationale was to:

- provide value for money and efficiency savings
- transfer risk from the public to the private sector

and these motives still remain pretty much the driving force as the procurement policy matures. In 2001, the IPPR Commission also came to the conclusion that PPPs are not a panacea for resolving all the challenges which modern government faces. They may be one important part of a wider strategy for the revitalisation of public service, but they are not a way of resolving how public services should be funded. In 2004, following *Building Better Partnerships*, the IPPR published a follow-up report entitled, *3 Steps Forward, 2 Steps Back – Reforming PPP Policy*. The 2004 report presented its conclusions stating that there is still confusion over the role of PPPs.

The PFI procurement process – 'Getting a good deal'

Table 4.4 illustrates the recommended Treasury Taskforce procurement path for PFI projects. The PricewaterhouseCoopers report *PFI Competence Framework* suggested these stages fall into three broad phases:

Table 4.4 PFI procurement guide (Source: HM Treasury)

Stage 1	*Establish business case.* It is vitally important that the PFI project is used to address pressing business needs. Consider key risks
Stage 2	*Appraise the options.* Identify and assess realistic alternative ways of achieving the business needs
Stage 3	*Outline business case.* Establish the project is affordable and 'PFIable'. A reference project or public sector comparator should be prepared to demonstrate value for money, including a quantification of key risks. Market soundings may be appropriate at this stage (see Chapter 6). The outline service specification should be prepared
Stage 4	*Developing the team.* Form procurement team with appropriate professional and negotiating skills
Stage 5	*Deciding tactics.* The nature and composition of the tender list and selection process
Stage 6	*Publish OJEU.* Contract notice published in OJEU (see Chapter 6 and Appendices)
Stage 7	*Prequalification of bidders.* Bidders need to demonstrate the ability to manage risk and deliver service
Stage 8	*Selection of bidders.* Short-list bidders. Method statements and technical details may be legitimately requested
Stage 9	*Refine the proposal.* Revisit original appraisal (stage 3) and refine the output specification, business case and public sector comparator
Stage 10	*Invitation to negotiate.* Could include draft contracts. Quite lengthy – 3 to 4 months. Opportunity for short-listed bidders to absorb contract criteria and respond with a formal bid
Stage 11	*Receipt and evaluation of bids.* Assessment of different proposals for service delivery
Stage 12	*Selection of preferred bidder.* Selection of preferred bidder with bid being tested against key criteria
Stage 13	*Contract award and financial close.* Sign contract and place contract award notice in OJEU (see Chapter 6)
Stage 14	*Contract management.* Operational and management relationship between public and private sectors

1. Phase (i), Feasibility – Stages 1–4
2. Phase (ii), Procurement – Stages 5–13
3. Phase (iii), Contract management – Stage 14.

Note: Although the Treasury Taskforce no longer exists, the data produced by this organisation is still recognised as being valid and appropriate. During the past 5 years or so, there have been various attempts to modify the procurement process as it has been criticised for lacking flexibility and being too long. For example, in the PFI project for the redevelopment of West Middlesex Hospital that opened in 2004, a round of bidding was omitted in order to speed up the process. Subsequently the National Audit Office concluded that the trust ran an effective bidding competition but that it should be noted that if this strategy was to be used in a future PFI deal then the following safeguards need to be put in place to maintain competitive tension when using this approach. It is recommended that the public sector client should:

- Obtain greater bid detail at an early stage
- Keep the main aspects of the deal constant in the closing stages
- Be prepared to walk away from the preferred bidder
- Make it clear to bidders that this process is to be applied
- Ensure that there are no major open issues for negotiation.

This section of the chapter will concentrate on the stages that have proven to be of crucial importance in determining the case for the use of PFI, namely:

- Stage 3 – the assessment of value for money and risk transfer, the Outline Business Case
- Stage 9 – Refining the proposal
- Stage 12 – Selection of the preferred bidder.

Stage 3 – The outline business case

The outline business case is at the heart of the feasibility phase. Assuming that the need for the project has been established and other methods of delivery considered, the preparation of the outline business case is the starting point of an audit trail that runs through the procurement process and should contain the following:

1. An identification of key risks
2. An output specification
3. A reference project or public sector comparator.

Identification of key risks

Risk transfer is one of the key tests for a good PFI deal as value for money can be demonstrated to increase each time a risk is transferred. There are two aspects to risk transfer:

1.   Transfer between the public and private sectors
2.   Transfer between the members of the PFI consortium.

In most PFI projects, the risks that are earmarked for transfer to the private sector are by now fairly standardised and well understood; however, major difficulties can arise in deciding who within the consortium carries the various burdens of risk. This factor would seem to explain why complete teams, that include both contractors and facilities management operators, are increasingly successful at winning PFI bids. In the case of risk transfer between public and private sectors, the main drivers are transparency and the need to demonstrate value for money, in the case of risk transfer within the consortium the commercial interests of the various players, i.e. financial institutions, contractors, operators, etc., dominate the discussion.

The principle governing risk transfer is that the risk should be allocated to whoever from the public or private sector is able to manage it at least cost, that is to say identified risks should either be retained, transferred or shared. The valuation of risk transfer, however problematic, often tips the scales on PFI deals as the public sector comparator alone often shows that value for money has not been demonstrated. In six of the 17 projects analysed in the Arthur Andersen/Enterprise LSE, the assessment of value for money was entirely dependent on risk transfer valuation and in all 17 projects it accounted for 60 per cent of all cost savings, making this element of the deal particularly important. There will always be a wide variety of risks associated with potential PFI projects, including the following:

Risk transfer between the public and private sectors

*   *Design and construction risk.* The construction period of a PFI project is recognised as one of the most critical phases and for this reason often attracts the highest valuation which in some cases can account for 50 per cent of all risk valuation. If for any reason the project is not complete or is late then the service, whether it is healthcare or education, cannot be delivered and the income stream will not be generated. Generally, the design and building of the asset is a risk best borne by the private sector consortium and its financial backers. Because the built asset is being designed to an output/service specification instead of a rigid set of departmental guidelines and there is a commercial

incentive for efficiency right through from initial design to build and operation.

- *Commissioning and operating risks* (including maintenance or whole-life costs). It has been estimated that over the 35-year life of a PFI contract, that on average 35 per cent of all costs will be capital cost, while 65 per cent will be running and maintenance costs. In fact, in some projects the split could be as great as 20/80 per cent. The golden rule therefore for consortia is 'concentrate on the larger'. As discussed in Chapter 1, the UK construction industry has traditionally ignored the influence of whole-life costs in building design; the PFI ensures that they are ignored at the consortium's peril. Once again this risk is best managed by the private sector.
- *Life cycle risks*. PFI contracts are long term and therefore SPCs have an incentive to fully investigate the impact of whole-life costs during the design phase of a project. Whole-life costs are thought by many to lie at the heart of PFI, particularly as stated earlier, when such a high percentage of costs associated with a PFI project are to be found in running and maintenance costs and every public sector comparator is built upon and judged on the basis of the net present value of whole-life costs. This makes sound commercial sense in the context of design, build, finance and operate, where risk is being transferred over a long time period and it has to be priced at the bid stage. Until comparatively recently, most buildings have been conceived and built on the basis of very simple criteria, fitness for purpose corresponding to the lowest possible construction cost. In addition, in many countries, including the UK, fiscal systems of taxation, tend to favour high running and maintenance costs over low capital costs. Whole-life costs are discussed in Chapter 3. The critical difference between PFI and typical private sector procurement is that the knowledge expertise and control over life cycle risks are in the hands of the service provider, who has a major incentive to optimise. Unless the life cycle risks are managed, the price for the job will be wrong. Often, neither the client nor the bidder has been able to pull together all of the necessary information.
- *Demand (volume) risk*. The prospect of receiving a stable long-term income flow that is a major attraction for many private sector consortia that bid for PFI projects. Therefore, demand or volume risk is uncertainty over the level of demand for the service provided by consortium-operated asset. Generally, the

private sector is unwilling to accept volume risk as usually it is the public sector that has control over volume – for example, the numbers of convicted prisoners requiring a place in custody or the numbers of patients requiring a hospital bed.

- *Residual value risk*. Residual value risk is uncertainty over what the net value of the asset will be at any time during the contract period. It is highlighted as an issue when a PFI scheme anticipates the transfer or sale of the asset and requires case-by-case consideration. For instance, the procuring entity may no longer require the asset at the end of the service contract. There are two main determinants of residual value; first, the condition of the asset at the end of the contract and second, the demand, if any, for the asset. Some projects, such as prisons, obviously limit the possible alternative uses for this type of project; however, the private sector take responsibility for maintaining the asset in good condition.

- *Technology and obsolescence risks*. Technology risk is associated with the obsolescence of both the services and the function of the assets themselves. It is generally not thought to be significant outside IT projects; however, it would be a brave person who tried to predict methods of healthcare delivery in 20 years time. Technology refreshment, for example replacing computer networks within a school every 5 or 10 years, is a risk usually transferred and managed by the private sector, but wider ranging obsolescence risks impacting on the mode of service delivery would be a matter for negotiation.

- *Legislation risks – both UK and EU*. Legislation or regulatory risk is one special aspect of demand risk and is thought to be outside the influence of the private sector. For example, a reduction in the resources available to the NHS due to legislation passed by either the UK Government, the Scottish Executive or the European Parliament. The test should be one of materiality, with the risk of general changes in the law being borne by the private sector although the way should be open to negotiate possible price adjustments in the event of changes with a major impact.

- *Project finance risk*. Project finance risk is the risk associated with the ability to raise finance on the terms suggested by the consortium in their bid. It is a risk retained by the private sector. One of the major criticisms of PFI was that the cost of using private sector finance would always be more expensive than public sector funding. The maturing of the PFI now means that the gap has narrowed.

- *Termination risk*. Most PFI contracts contain a statement of the circumstances that would result in the contract being terminated. Generally, termination occurs if the consortia fail to provide the level of performance statement in the service specification. Under these circumstances, the public sector would be exposed to taking over the running of the service and paying off any debt associated with the project.
- *Refinancing risk*. The National Audit Office now recommends that risks associated with the refinancing of PFI deals is considered at an early stage. Refinancing, which is explained below, can result in the public sector being exposed to an increase of termination risk – that is additional debt that has to be repaid in the case of the contract with the private company being terminated.

Refinancing and clawback

One of the more controversial aspects of PPP deals and in particular the early PFI deals is the ability of the private sector consortia to renegotiate their debt during the currency of the contract; a process referred to as refinancing. The practice first sprung into the PPP headlines in 1999 after the details were made known of the refinancing of any early PFI deal for Frazakerley prison (now known as Altcourse) near Liverpool that resulted in gains of around £10.7 million for the SPC, an increase of 61 per cent on original calculations. At the time of the original deal in December 1995 there was no provision in the contract to cover this process except to seek the permission of the Prison Service and only £1 million of the proceeds of the refinancing was given to the public sector. The refinancing process enabled early repayment of debt and the generation of high dividends for share/equity holders. The refinancing of Altcourse has four strands:

1. An extension to the period over which the SPC, in this case Frazakerley Prison Services Ltd (FPSL), bank loan would be repaid
2. A reduction in the lending margin for the loan
3. An arrangement of a fixed rate of interest of the full period of the loan
4. Early repayment of the subordinated debt invested by shareholders of FPSL.

Other factors particular to this contract were the completion of the facility ahead of schedule and lower construction and commissioning cost than originally allowed for, which, taken in association with the refinancing costs, produced a total increase in shareholders' expected returns of £14.1 million, a total of 81 per cent higher than when the contract was awarded.

As a direct result of the Frazakerley deal, the National Audit Office published a report in June 2000 setting out general principles which government departments should apply to refinancing:

- Appropriate benefits should go to those bearing risks
- Benefits from reducing costs in a developing market should be shared if they have not previously been reflected in the contract price
- Departments that sponsor refinanced projects should seek compensation for any exposure to increased risk
- Substantial refinancing gains may threaten public perception of value for money
- If the private sector seeks refinancing then it is reasonable for the public sector to seek a share of the benefits.

Refinancing is still common place for PFI projects, but the degree of transfer of other risks is a matter for negotiation at project level. In addition to the NAO's report, in July 2001, the Office of Government Commerce issued guidance for government departments when refinancing PFI projects in which it urged caution on departments and suggested that, with regard to any reallocation of risk, any benefits from refinancing should be shared 50/50 with the private sector partner.

Assessing risk

The value of risk transfer, both quantitative and qualitative, is a vital part in the assessment of VFM in the PSC of a PFI project. The process involves the following stages:

1. Identification of risks
2. Assessment of the impact of risks
3. Assessment of the likelihood of risks arising, including adjustment of the financial model
4. Allocation of risks.

For a large PFI project, the identification of risk is likely to be a long and complex procedure and the interaction between various

Table 4.5 The risk register

Commercial risk	Purchaser	Operator
Demand risk	(£10 m)	
Third party revenues		£20 m
Design risk		(£25 m)
Maintenance risk		(£20 m)
Obsolescence		(£25 m)
Residual value	£10 m	

identified risks must also be considered. The allocation of risk with the preferred bidder will be the subject of negotiation; however, for the purpose of establishing the feasibility of the project and the Public Sector Comparator (PSC) the following assumptions have been made in this case, based on previous experience.

- *Stage 1 – allocate risk.* Table 4.5 lists some of the risks that have been identified for a prison project together with an initial net present value valuation the figures in brackets represent possible losses.
- *Stage 2 – Estimate commercial significance of risk.* The figures in the risk register are drawn either from empirical evidence or where that is not available from common sense estimates based on specialist knowledge. As the assessment of risk transfer is such an important part of proving value for money the figures should be as accurate and as up to date as possible.
- *Stage 3 – Assessing the likelihood of risks arising.* Having established the value of the risk in net present value terms, the next step is to calculate the probability that everything will go according to plan and for example the demand risk is correctly valued at £10 million. The most accurate and widely accepted method of doing this is by using a simulation modelling technique, of which there are many, but perhaps Monte Carlo simulation is the most respected. In order to apply this technique it will be necessary to prepare an estimate of the probability distribution within upper and low limits. After applying a Monte Carlo simulation to the demand risk estimate, it is shown that there is only a 70 per cent probability that demand for the new facility will decrease to the point where the purchaser would lose money, therefore the estimate can be downgraded to £7 million. The quality and accuracy of the information used to

construct the simulation model is of crucial importance. This exercise is then repeated for every risk identified by the purchaser and the financial model is adjusted.

- *Stage 4 – Allocation of risk.* Who is now going to accept the risk? The approach taken by some purchasers to allocation of risk is to issue a risk matrix to the short-listed consortia at stages 7/8 in the procurement process, marking those items that are clearly not negotiable and are to be transferred, but requesting short-listed consortia to indicate which on the remaining risks they are prepared to accept.

Risk transfer within the consortium

As discussed previously in the chapter, a consortium is a unique collection of organisations with very differing expertise, each attempting to achieve maximum return combined with minimum exposure to risk. One scenario for trying to achieve this is with the use of a financial model into which factors such as desired return, whole-life costs, capital costs, generally based on costs/m² of gross floor area, are input, thereby determining what's left to compensate for the risk. This figure is then divided after considerable debate among the various consortium members.

Output specification

The preparation of the output specification is one of the most important phases in the PFI bidding process. It is vital that the output specification states the core requirements in terms of output, that is to say what is wanted, not how it is to be provided, thereby allowing the consortia the maximum opportunity for innovation in service delivery. The key elements of an output specification are seen as:

1. *Objectives.* A clear statement of the strategic objectives of the project expressed if possible in terms of delivery of service rather than the built assets.
2. *Purpose.* A summary of the desired outputs of the project, for example a new PFI hospital project, may describe its purpose as follows:

   This project is to provide healthcare services for:
   - Thirty long-stay beds for elderly people
   - Nineteen acute GP beds

- Out-patient services, including specialist input
- Day surgery and minor injury clinic.

The service will include the provision of catering, cleaning, hairdressing, transport and porterage (including ambulances), and security, together with arrangements for the maintenance of patient records.

3. *Scope.* The purpose of this section is to identify whether the proposed project has the potential to produce alternative income streams. For example, could a school building be used during weekends, holidays, etc., or could a hospital be extended to include the provision of private accommodation, and who is to benefit from any extra income stream that this additional use may attract. Market sounding may be appropriate to ascertain the potential for the type of action. Clawback clauses are inserted into PFI contracts to share extra profits between the operator and the public sector clients. When no such clauses exist, the PFI consortium share out the profit among themselves, which some argue is justified given the risks involved. For example, when the Special Purpose Company restructured their debt for Frazakerley Prison after the financial close it resulted in a near doubling of profit from 16.5 million to 30.6 million. Government guidelines are expected to be issued covering this contentious issue.

4. *Performance.* The required performance levels of the project should be set out by way of operating outputs without reference as to how this performance will be achieved. Importantly, the performance of the operator should be capable of being measured and evaluated, as payment will be directly related to this. Government agencies like the National Health Service and the Prison Service have standard criteria in areas such as waiting times for hospital admission, facilities and timing of recreation periods within a prison, etc. and these should be maintained by the private sector operator as a minimum level of service provision. One of the clouds hanging over the PFI is the issue of what happens if an operator fails to deliver a service to the required levels, as has happened in at least one case, when the public sector had to step in to take over the service delivery.

5. *Compliance and compatibility.* In the situation where a private operator is to provide a service within the framework of a larger service delivery, it is fundamental that the service being

provided under the PFI project is compatible with the overall systems. One obvious example is the formats of IT systems.

6. *Constraints*. If the project is the subject of constraints, for example in terms of planning permission, then these should be made explicit.

7. *Risks*. The type and nature of risks should be set out as illustrated in Table 4.5.

8. *Alternative solutions*. Allow tenderers the opportunity to offer alternative solutions, without being prescriptive as to the nature of what these alternatives may be, for example: 'Suppliers will be responsible for the maintenance or replacement of electrical and mechanical equipment throughout the life of the contract to a standard that permits service standards to be met'.

Most PFI deals currently being negotiated will run for 30 years plus and one of the major problems to be addressed when considering outputs are the ways in which demands on the service may alter during that period. For example it is particularly difficult to plan for healthcare delivery against the back drop of an increasingly ageing population, the number of people in the UK over 85 increased threefold between 1971 and 2000, from 400 000 to 1.2 million and the increasing expectations of a consumer society for high quality care. In addition, there is the emergence of new technology, surgical practices, diseases and cures. In order to allow for change, purchasers should specify that the design of assets allows for some operating flexibility and contracts should permit changes to the unitary charge arising from service changes.

Reference project or public sector comparator

Public sector comparators (PSC) incorporate a public sector reference project that provides a snapshot of value for money at a particular point in time. Put simply a PSC poses the question – which gives better value for money – the conventional public procurement route or PFI? The lack of transparency in the construction of a PSC has been criticised and indeed there has been the accusation that figures that are used in this pivotal VFM exercise have not been scrutinised sufficiently rigorously. A PSC, or reference project, may be defined as a hypothetical risk-adjusted costing, by the public sector as a supplier, to an output specification produced as part of a PFI procurement exercise and

- Is expressed in net present value terms
- Is based on the recent actual public sector method of providing that defined output (including any reasonably foreseeable efficiencies the public sector could make)
- Takes full account of the risks which would be encountered by that style of procurement.

In order to be a valid benchmark against which private sector bids can be fully and fairly compared, the PSC must not only reflect certain procurement costs but also the additional costs of risks that may arise, which under PFI would fall to the consortium. During the PSC process, risks should be identified and the cost impact evaluated.

The format of a PSC is not fixed. However, the *Treasury Taskforce's Technical Note No. 5, How to Construct a Public Sector Comparator*, does set out a recommended list of items that should be included. These are:

- An overview and short description of the project
- An estimate of basic procurement costs and operating costs
- The opportunities for third party revenues
- An estimate of the value of the asset on transfer or disposal
- A risk matrix showing their costs and consequences
- A discounted cash flow of future costs
- A sensitivity analysis showing the consequences of variance in key assumptions.

Stage 9 – Revisit and refine the original appraisal

The outline business case is now complete and assuming that VFM has been established the procurement procedure can proceed. The figures prepared in stage 3 will be revisited at stage 9 before selected bidders are asked to submit their proposals and enter into negotiations. The original output specification should be refined to a set of core deliverable outputs and a set of proposed key contractual terms and allocation of project risks prepared. Having selected a short list of prequalified bidders earlier in the process, the documentation is now ready to distribute during stage 10.

Stage 12 – Selection of preferred bidder and final evaluation

It is not uncommon for stages 10 and 11 to take several months. On selection of the preferred bidder, the PFI project should again be tested against the critical success factors of value for money including comparing the preferred bid with the public sector comparator. Concern has been expressed regarding the value of the PSC, as it is felt by some departments that, as PFI is the only procurement path likely to get approval, the degree of scrutiny of the figures used to compile it could be more rigorous. In an attempt to introduce more objectivity into the process, from January 2001 the Gateway Process (see Figure 4.6) has been mandatory for all new high-risk procurement projects and all IT procurement projects in

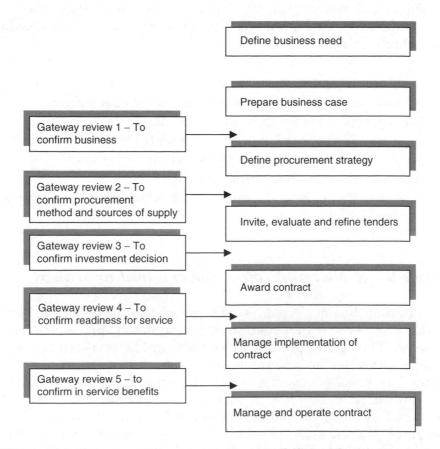

Figure 4.6 Gateway review process (Source: Office of Government Commerce).

Table 4.6 PFI credits

Year	Payments to the project company	Credit – NPV
1	£0	£0
2	£0	£0
3	£666 667	£572 619
4	£1 007 500	£802 012
5	£1 015 056	£748 866
6	£1 022 669	£699 243
	↓	
22	£1 152 540	£233 456

central government, executive agencies and other non-departmental public bodies. The Office for Government Commerce introduced the procedure in order to apply an independent assessment and further scrutiny of the project being considered. The process is carried out by a team of between two and five people, depending on the nature of the project, and there can be up to five reviews during the life of a project, usually taking place over a 3–5-day period. The recommended timing of the reviews is shown in Figure 4.6.

Indeed a key feature of the PFI process is that it takes a long time to arrive at an agreed deal. Figures produced by the Department of Health indicate that from the publication of the OJEU notice to financial close was 46 months, while the Private Finance Unit of the Scottish Executive estimate that for PFI schools projects the period is between 98 and 116 weeks. PFI projects are complex, they are not a quick fix to an investment need, but the process should become quicker and parties become more familiar with this form of procurement.

Public private partnership projects (4Ps)

This form of PPP was introduced in 1996 with the intention of encouraging local authorities and councils to consider PPPs for the delivery of some services. 4Ps projects are long-term contracts, 30 years plus, and are really a form of PFI in which central government through various departments, such as the Office of the Deputy Prime Minister, subsidises local authorities by helping them to pay the unitary charge with what are known as, PFI credits. In 2005, the amount of credits or subsidies available were;

Department of Health – £70 million
Office of the Deputy Prime Minister – £100 million
Department for Culture, Media and Sport – £130 million.

The minimum limit for consideration is projects with an initial capital element of at least £20 million. Obviously the demand for this source of funding exceeds the supply, therefore local authorities are required to submit bids, with funding only being awarded to approximately 20 projects. As the credits will only fund part of the cost of the project, the local authorities must secure other forms of income streams. Each of the submissions is subjected to a two part assessment where projects are judged against a set of criteria which changes from department to department and year to year. For example, priority for submissions to the Department for Culture, Media and Sport for the period 2006–08 would be projects that promote:

- the modernisation of the library service;
- the creation of multi-sport facilities.

Projects may seek approval from more than one government department and more specifically will be expected to address issues such as social inclusion, innovation in service delivery and long-term flexibility.

As in PFI projects, applications have to be supported by robust business cases that clearly demonstrate value for money and evidence of need and demand. Table 4.6 illustrates how the PFI credit systems works. No payment is made until the facility is operational, assumed to be 2 years. In year 1, the facility is completed and payments to the project company commence at a rate of £666 667 for which the local authority receives a credit, calculated on a net present value basis as £572 619. As can be seen, payments then continue over the life of the contract, 22 years in this case, but while the payments to the project company remains fairly constant the credit decreases.

The role of the quantity surveyor in PPP/PFI

Quantity surveyors are just one of a range of professional advisors involved in PPP/PFI projects. As previously noted, the Royal Institution of Chartered Surveyors are in no doubt as to the importance of PPP/PFI in the future of the profession and many quantity surveying practices are involved in PPP/PFI deals in a variety of roles for both the public and private sectors. Figure 4.7 illustrates the public/private skills balance as identified by the RICS Project Management Faculty. In the private sector, working

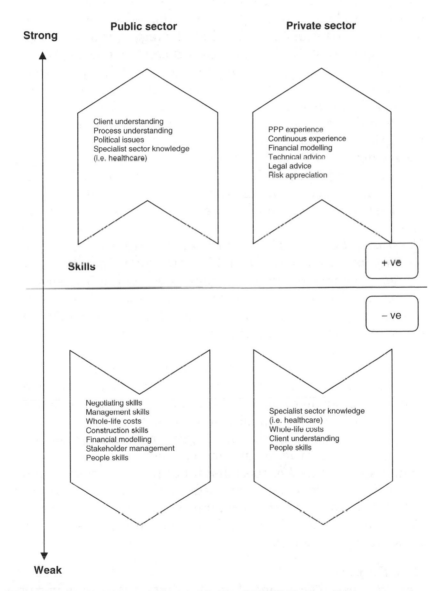

Figure 4.7 PPP skills balance (Source: RICS Project Management Faculty).

for the operator, the quantity surveyor role could involve (note that all references are to Table 4.4):

1. *For the private sector – special purpose companies:*
- *Advice on procurement.* For many private sector consortia, the approach to submitting a bid for a PFI project is unknown territory. Added to which is the fact that by their nature PFI projects tend to be highly complex, requiring decisions to be taken during the development of the bid at Stage 8, that not only involve capital costs, but also long-term costs. Increasingly, as explained earlier in this chapter, the impact of EU procurement directives must be considered. Some contracting authorities use the OJEU to 'test the water' for a proposed PPP project at Stage 6; the quantity surveyor can supply preliminary cost information at this time. In addition the quantity surveyor with experience in PPP can provide expert input into the pre-qualification stage – Stage 7. The stage at which the bidders are selected, based upon (among other things) their knowledge of a specialised sector of public services and their ability to manage risk.
- *General cost advice.* The traditional quantity surveying role of advising on capital costs, including also the preparation of preliminary estimates, bills of quantities, obtaining specialist quotes, etc. In addition, value management and value engineering techniques described in Chapter 2 are increasingly being called on to produce cost-effective design solutions.
- *Reviewing bids prior to submission* – due diligence.
- *Advice on whole-life costs.* It has already been stated that to many the key to running a successful PPP contract is control of whole-life costs. Recognising this many surveying practices now have in-house advice available in this field.
- *Specialist advice.* Obviously, highly complex projects, for example the construction and management of a major hospital, requires a great deal of specialist input from, for example, medical planners able to advise on medical equipment, etc. from the outset. Surveying practices committed to developing their role in PPP already have at their disposal such expertise, which in some cases is in house.

2. *For funders:*
- *Due diligence.* The financial and funding aspects of major projects are becoming increasingly susceptible to both technical, as

well as commercial risks. Investors and funding institutions are looking more and more for independent scruniy of all aspects of development from design integrity to contractual robustness of the contract and beyond to the expenditure levels and progress against programme. The skills of the quantity surveyor provide an excellent platform for the investigative and analytical processes necessary to satisfy these requirements.

3. **For the public sector purchaser:**
*   *Procurement advice.* This method of procurement is for many, just as unfamiliar as the private sector. The surveyor can advise the contracting authority on how to satisfy the requirement of this method of procurement. It is widely agreed that the appointment of a project manager at an early stage is vital to PPP project success. In addition, pressure is being exerted to speed up the procurement process – a factor that makes the role of the project manager even more crucial.
*   *The outline business case.* The preparation and development of the OBC in Stage 3 involves the preparation of a risk register, the identification and quantification of risk; all of which are services that can be supplied by the quantity surveyor.
*   *Advice on facilities management.* Technical advice on this topic during the drawing up of the service specification, at Stage 3 and beyond.

4. **Common and joint services to SPCs/public sector purchasers – joint public/private monitor certifier.** This role is similar to the role played by a bureau de contrôle in France and involves monitoring the construction works to ensure that it complies with contract. In addition to the built asset, the surveyor employed in this role can monitor facilities management operation. The concerns with this practice centre around the 'belt and braces' way in which the certification is being carried out and the fact that firms are signing off multi-million pound schemes, for very little fee and are effectively acting as unpaid insurance agents, with any claim being covered by professional indemnity insurance.
5. **Services to consortium building contractors.** The role recognised by many surveyors as the main involvement in PPP. It includes preliminary cost advice, preparation and pricing of bills of quantities, and supply chain management.

Emerging trends in PPP procurement

1. *Standardisation of contract conditions.* Many of the consortia now bidding for PPP/PFI projects are experienced in the procurement procedure to the extent that increasing standard terms and conditions are being developed in order to speed up the procurement process.
2. *Mixed sector projects.* The provision of services to housing, education and healthcare, for example, could be combined in the single deal, where the value of the individual deal is too small to make PPP possible.
3. *Bundling.* Bundling together several smaller PFI schemes that would not have critical financial mass individually.

Conclusions

The UK Government's preferred methods of construction procurement have been identified as:

*   PPPs
*   Prime contracting
*   Design and build.

Therefore, it is safe to assume that PPPs and the PFI are here to stay in the medium term. Over recent years the amount of adverse media publicity, particularly for PFI projects, has seemed to diminish, but why this is so is hard to determine. Perhaps it is now perceived that the PFI is beginning to deliver new schools and hospitals or it could be that there has been an acceptance of the inevitable. As far as the UK construction industry is concerned, PPPs have proved to be a very profitable venture – with a few notable exceptions – producing higher profits than traditional contracting. PPPs, as has been described, are also developing and diversifying into specialist models for specific sectors and this trend will continue. Such is the concern of the motives and effectiveness of the PFI that the NAO has produced around 37 reports both on specific PFI projects, as well as themes relating to the PFI. In fact when compared with other forms of government spending PFI projects have had a disproportionate number of reports/audits. Speaking at the MPA's 21st annual conference, Sir John Brown, Comptroller and Auditor General of the UK and Head of the National Audit

Office stated, 'Yes the PFI and PPP projects are working and are the vehicle for the delivery of public services in the 21st century'. Of course 'working' means different things to different groups of people and in a relatively immature marketplace characterised by long-term projects, there is clearly difficulty in trying to assess the success of the PFI to live up to its aims of delivering value for money and efficient public services. Fortunately, over the recent past there have been many reports by independent organisations such as the National Audit Office, Audit Scotland and the Henley Research Study into the performance of PFI projects. One area of justified criticism is the continued lack of transparency surrounding PFI deals. Concerned about the lack of transparency in some PPP deals, the Institute for Public Policy Research published the results of its Openness Survey in February 2004 which concentrated on the availability of information on PPP projects. The report concluded that PPPs disrupt traditional accountability structures; however, they can open up new routes of accountability provided that certain standards are complied with. One of these standards is openness which sits at the base of many accountability considerations. The report further concluded that to date openness considerations have not been fully resolved and as a result many people are still suspicious about such items as value for money. On the other hand, the IPPR did recognise that some information should remain confidential and needs to be kept secret in order to render projects viable. Typically these include the need to keep trade secrets confidential and to safeguard the competitive positions of the public and private sectors. However, the report concluded that it is important to ensure that commercial confidentiality is not employed as a spurious justification to withhold information from the public domain.

The international development of PPP

PPPs are being adopted by governments worldwide with European PPPs, including the UK, accounting for 85 per cent of all PPP contracts. This worldwide trend has opened up many opportunities for UK quantity surveyors (Figure 4.8).

Even where there is strong political motivation to develop PPPs, the complexity of the individual projects/contracts and the need to develop the capability to create an enabling environment, results in progress being slow initially. In countries where PPPs have been

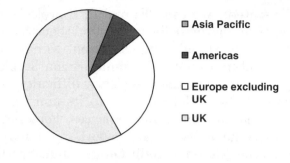

Capital value of closed deals in 2000

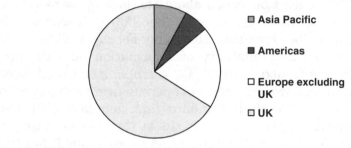

Capital value of closed deals in 2003

Figure 4.8 Capital value of PPP deals (Source: Dealogic).

introduced, generally speaking, they have developed along the following lines.

A highly centralised 'big picture' model as developed in Ontario, Canada. The Ontario SuperBuild Corporation, effectively a private company, was created in December 1999 in response to a report by the Ontario Jobs Investment Board in which the board called attention to the province's severe infrastructure decline. Many of the schools and hospitals in Ontario date back to the 1920s, whereas roads and sewerage treatment plants are now over 50 years old. The scenario is a classic one. Billions of dollars of investment needed and new projects required to come on stream but without the traditional long lead in times and unpredictable completion dates. The government responded by creating the SuperBuild Corporation as a $20 billion, 5-year initiative, to address Ontario's infrastructure needs and meet the economic challenges of the new millennium. The intention was that the initiative's goals would be achieved through the use of PPPs. SuperBuild's mission was to co-ordinate strategic planning and capital expenditure which in the past had been dealt with by each provincial ministry, hospitals, education, etc., who received their funding. Ministries are required to take stock of the age, condition and value of capital assets, as well as projecting future capital needs. Hospitals and universities must also develop long-term capital asset plans as a condition for receiving capital investment support. Approvals for capital plans by ministries must be referred to SuperBuild and then to the Cabinet Committee on Privatisation which undertakes a strategic review and develops a capital plan for the government. In May 2005, the SuperBuild Corporation was rebadged the Ontario Infrastructure Project Corporation with a focus on non-traditional procurement strategies. Perhaps being conscious of the poor image of PPPs in certain parts of the media, the government decided to rebadge PPPs as Alternative Finance and Procurement, stating the drive to renew public sector projects would be based on a set of policies to guide private sector participation in public sector projects entitled, *'Building a Better Tomorrow'*. This framework establishes very clear requirements that must be met at every step of any public infrastructure project where the government has a financial interest. The framework now applies to every major project in Ontario like the North–South Light Rail Transit Project in Ottawa, the Durham Courthouse, and the North Bay Hospital. The principles for the new initiative are based upon protection of the public interest, which is to be achieved in the following ways:

- Projects will be undertaken only if they will improve the delivery of public services or contribute to the province's economic competitiveness.
- Appropriate public ownership and control over our core public services. Core services like hospitals will always be publicly owned.
- The third is value for money. Projects will only get the green light if a strong business case can be made to show that they will benefit taxpayers.
- Accountability. Stakeholders, both the public and private sectors, will be held accountable for delivering the project. Projects will be subject to review by an independent third party.
- Finally, fairness, efficiency and transparency.

A somewhat different approach is the decentralised model which has been developed in Portugal where PPPs are becoming an important element of infrastructure with a number of road and water treatment projects. However, despite the increased use of PPPs, no specific structures have been put in place by the government to assist and co-ordinate delivery of projects. This has very much been left to individual government departments and local authorities. The Portuguese government has established an informal task force to provide advice on forms of PPPs, legislative barriers, etc. The disadvantages of this approach are the necessity to reinvent the wheel for every new PPP project resulting in lengthy and expensive procurement processes and the loss of expertise gained during projects.

Finally, the most common model is a mix of centralised and decentralised approaches, as exemplified by models developed in the UK and the Republic of Ireland. In these models there exists a central co-ordinating and policy development units mainly focused on project development and procurement. The unit has no role in project delivery.

Bibliography

Audit Commission (2001). *Building for the Future. The Management of Procurement under the Private Finance Initiative*. HMSO, London.

Buckley, C. (1996). Clarke and CBI unite to revive PFI, *The Times* 22 October.

Building (2001). PFI plan to liberate architects, 31 August.

Cartlidge, D. (2006). *Public Private Partnerships in Construction*. Taylor and Francis, Oxford.

Catalyst Trust (2001). *A Response to the IPPR Commission on PPPs*. Central Books, London.

CGR Technical Note 1 (2004). HM Treasury.

Ernst & Young (2003). *PFI Grows Up*.

EU Commission (2003). *Guidelines for Successful Public Private Partnerships*, European Commission, Brussels.

EU Commission (2004). *On Public Private Partnerships and Community Law on Public Contracts and Concessions*, European Commission, Brussels.

Gosling, T. (ed.) (2004). *3 Steps Forward, 2 Steps Back*, Institute for Public Policy Research.

HM Treasury (1998). *Stability and Investment for the Long Term*, Economic and Fiscal Strategy Report – Cm 3978 (June), HMSO, London.

HM Treasury (1998). *Policy Statement No. 2 – Public Sector Comparators and Value for Money*, HMSO, London.

HM Treasury (1999). *Technical Note No. 5 – How to Construct a Public Sector Comparator*, HMSO, London.

Institute for Public Policy Research (2001). *Management Paper, Building Better Partnerships*, London.

Kelly, G. (2000). *The New Partnership Agenda*, The Institute of Public Policy Research, London.

Meeting the Investment Challenge (2003). Office of Government Commerce.

Ministry of Finance (2000). *Public–Private Partnership – Pulling Together*, PPP Knowledge Centre, The Hague.

National Audit Office (1999). *Examining the Value for Money of Deals under the Private Finance Initiative*, HMSO, London.

National Audit Office (2001). *Innovation in PFI Financing, The Treasury Building Project*, HMSO, London.

National Audit Office (2001). *Managing the Relationship to Secure a Successful Partnership in PFI Projects*. HMSO, London.

Office of the Deputy Prime Minister (2004). *Local Authority Private Finance Initiative: Proposals for New Projects*.

Partnerships for Schools (2004). *Building Schools for the Future*.

PricewaterhouseCooper (1999). *PFI Competence Framework – Version 1*, (Dec), Private Finance Unit, HMSO, London.

PricewaterhouseCooper (2001). *Public Private Partnerships: A Clearer View*, Oct, London.

4Ps (2000). *Calculating the PFI Credit and Revenue Support for Local Authority PFI Schemes*, 4Ps Guidance, March. HMSO, London.

Robinson, P. *et al.* (2000). *The Private Finance Initiative – Saviour, Villain or Irrelevance*, The Institute of Public Policy Research, London.

Robinson, P. (2001). *PPP Tips the Balance*, Public Service Review PFI/PPP, Public Service Communication Agency Ltd, UK. ISSN 1471-6046.

Rose, N. (2001). Challenges to procurement: the IPPR and Byatt Reports, *Government Opportunities* 24 July.

Thomas, R. (1996). Initiative fails the test of viability, *Guardian* 22 October.

Wagstaff, M. (2005). The case against, *Building Magazine* 14 January.

Waites, C. (2001). *Are We Really Getting Value for Money?* Public Service Review PFI/PPP 2001, Public Service Communication Agency Ltd, UK. ISSN 1471-6046.

Williams, B. (2001). *EU Facilities Economics*, BEB Ltd.

Zitron, J. (2004). *PFI and PPP: Client and Practitioner Perspectives*. Proceedings of 21st Annual Conference of the Major Projects Association, Major Project Association.

Websites

www.dfee.gov.uk
www.doh.gov.uk
www.hm-treasury.gov.uk
www.minfin.nl/pps
www.nao.gov.uk
www.mod.gov.uk
www.ppp.gov.ie
www.4Ps.co.uk

5

Procurement – doing deals electronically

Introduction

The chapter continues with the theme of doing deals, by examining the changing environment and nature of procurement in general and the impact of electronic commerce on quantity surveying practice in particular.

At the time of writing the first edition of this book, it seemed as though, in common with most other market sectors, construction including quantity surveying, was on the verge of an electronic revolution. There were exaggerated predictions from practically every quarter, including government departments, on the ways in which e-commerce and e-construction were going to become all pervading and revolutionise everything from tendering to project management. The reality, as described in the following paragraphs, has been somewhat different! Nevertheless, the perception of e-commerce by construction related organisations and professions is mostly positive, although to date objective measures have been missing to accurately benchmark the spread of e-construction. There is now a universal consensus that reliable e-commerce metrics are needed to track developments in this area and to understand its impact on our economies and societies. The OECD identifies the research and measurement priorities as depicted in Table 5.1.

e-Commerce defined

e-Commerce is a major business innovation which tends to be successful when led by commercial rather than technological considerations. e-Commerce exploits information and communication technologies (ICTs) to re-engineer processes along an

Table 5.1 OECD research and measurement priorities for e-commerce

Readiness	Potential usage and access Technology infrastructure/socioeconomic infrastructure
Intensity	Transaction/business size Nature of transaction/business
Impact	Efficiency gains Employment/skill composition; work organisation New products/services New business models Contribution to wealth creations Changes in product/value chains

organisation's value chain in order to lower costs, improve efficiency and productivity, shorten lead-in times and provide better customer service. Electronic commerce or e-commerce therefore consists of the buying, selling, marketing, and servicing of products or services over computer networks. Strictly speaking, according to the OECD, e-commerce may be defined only as the method by which an order is placed or received and does not extend to the method of payment or channel of delivery. Therefore, for the purist, the simple process of gathering information on line or sending an e-mail does not constitute e-commerce. Whatever the arguments over the definition, for the quantity surveyor, e-commerce allows instant communications through the supply chain giving the partners a clear real-time picture of supply and demand.

One of the first organisations to use the term e-commerce was IBM when, in October 1997, it launched a thematic campaign built around the term. Over the past few years major corporations have re-engineered their business in terms of the internet and its new culture and capabilities and construction and surveying have started to follow this trend. As with any new innovation there are forces that act to drive forward the new ideas, as well as those forces that act as inhibitors to progress and these are illustrated in Figure 5.1 and discussed later in the chapter. e-Commerce sprang to public attention in 1997, after the meteoric rise in the value of the so-called dot.com companies. Many within the construction industry were sceptical about the application of e-commerce to construction and in 2000/2001 when the value of shares on the UK stock market collapsed and many virtual companies evaporated over night, leaving massive debts and red faces, there was a collective; 'I told you so!'

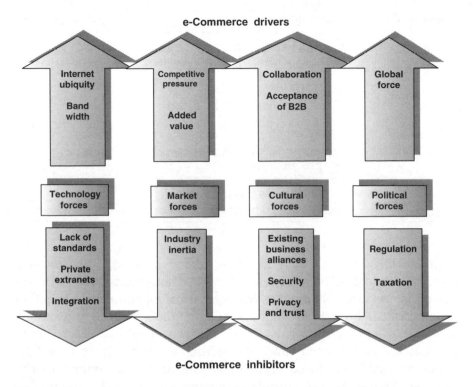

Figure 5.1 e-Commerce drivers and inhibitors (Source: Gartner Group).

The general consensus is that the failure of many dot.com companies occurred because of a combination of some or all of the following factors:

- Poor business models
- Poor management
- Aggressive spending
- Forgetting the customer
- The changing attitudes of venture capitalists. In 1999/2000 alone it was estimated that the total amount of venture capital investment in B2B (see later definition) exchanges was $25 billion worldwide, but this decreased to a trickle by 2001.

In common with other market sectors, e-construction has had a number of false dawns. For example, at the height of dot.com mania, five of the largest contractors in the UK announced the creation of the first industry wide electronic marketplace offering; buying building materials online, as well as a project collaboration package. Just over a year later it was announced that the planned internet portal was to be shelved, due to lack of interest. Despite this lack of enthusiasm from within the industry and the professions, due in large part to the continuing client-led drive for efficiency and added value, as well as examples from other sectors, the move to become an e-enabled industry continues. Information technology is at the heart of the developing tools and technologies that pool information into databases. An equally important aspect is looking at the attitudes of the people who need to feed information into and use the system and this will be discussed in detail in Chapter 6. The internet in particular provides a platform for changing relationships between clients, surveyors, contractors and suppliers; open exchange of information is critical in order to harness the best from this virtual marketplace and one of the biggest challenges is to create a culture that encourages and rewards the sharing of information. Too many people are still starting from the viewpoint that knowledge is power and commercial advantage, and the belief that the more that they keep the knowledge to themselves, the more they will be protecting their power and position.

Despite the somewhat slower uptake by the construction industry compared with other sectors, Figure 5.2 illustrates the wide range of e-markets available for construction and quantity surveying applications from basic e-shop websites to complex value chain integrators and collaboration platforms.

Until recently, business on the internet has been dominated by technology driven companies selling well-financed ideas for, as discussed previously, start up dot.coms which enjoyed the luxury of abundant investment capital, without the burden of having to show a profit. This honeymoon period has, however, come to an abrupt end. Though initially they lagged behind the 'idea companies', traditional 'bricks and mortar' companies are quietly making up for lost time and going on line. The current wisdom holds that it will likely be these companies that will ultimately become the financial cornerstones of e-commerce.

Why then has e-commerce not become the dominant way of conducting business in the construction sector? In a survey of the construction industry commissioned by the Construction Products Association in 2005, 83 per cent of those surveyed cited the culture of the industry as the major constraint on the development of e-commerce in construction. The survey also concluded that at first sight little had changed since 2000 when 86 per cent of those

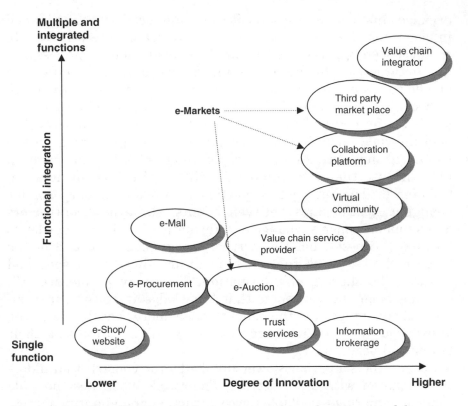

Figure 5.2 Classification of electronic commerce business models (Source: Paul Timmers).

questioned in a similar survey cited the same cause for the slow up-take of e-commerce. However, despite this negativity, 80 per cent believed that the construction industry is still committed to embracing new technologies. However, growth at 26 per cent had been much slower than the predicted 50 per cent in 2001.

The key advantages for the adoption of e-technologies were considered, by respondents in the 2005 survey, to be:

- Reducing costs
- Faster transactions
- Fewer errors
- Less paperwork.

Interestingly, 'access to new markets' which ranked very highly in the early days of e-commerce surveys has now disappeared from the list of perceived advantages.

The main disadvantages of e-technologies were seen to be:

- Initial set-up costs
- Loss of personal contact
- Retraining of staff.

One of the more surprising results of the Construction Products Association survey was respondents' views on the effect of e-commerce on construction products. Whilst 80 per cent of the industry as a whole and 69 per cent of manufacturers and distributors expected little or no change in the range of products available, 10 per cent of the industry as a whole and 19 per cent of construction product manufacturers expected a decrease on product awareness. In looking forward to 2010, when asked to predict how the proportion of business done with their suppliers and customers using e-commerce would change, the highest increase was envisaged to be with suppliers, with an overall increase of 40 per cent compared with 30 per cent with customers. The figures above indicate that to date, the majority of organisations would seem to utilise e-commerce to track the competition and improve communications. In addition it would seem that many companies, particularly SMEs, are engaging in e-commerce activities as a result of competitive pressure, suggesting a defensive line of action, rather than a differentiated one. However, as clients become more e-enabled, quantity surveying practices must follow or be left behind. As in the case of supply chain management techniques, and certain public sector procurement agencies (see Chapter 2) the pressure will come from the client, but even so there are still few, if any, quantity surveying practices which have made the leap from simple website to transaction platforms.

A high percentage of quantity surveying practices now have their own website. Software such as Microsoft FrontPage, as well as inexpensive proprietary website templates make the process of producing a professional looking website comparatively simple and inexpensive. This so-called first generation presence is used mainly for marketing and utilises the simplest form of business model illustrated in Figure 5.2. The so-called second and third generation presence, which incorporates transaction applications, have shown a much slower growth rate, particularly in the construction sector.

In a survey of e-commerce adoption carried out in 2003 by Statistical Indicators Benchmarking the Information Society

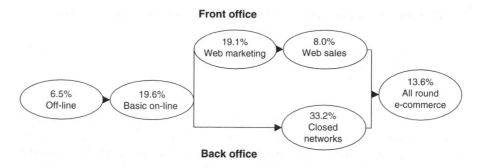

Figure 5.3 e-Commerce development stages (Source: empirica/SIBIS).

(SIBIS) across seven EU states, it was found that the leading e-commerce all-rounders were in distribution and financial network sectors where as off-line and basic on-line organisations were most likely to be found in construction and manufacturing.

As shown in Figure 5.3, internet technologies can be exploited in marketing and sales by introducing web marketing and eventually e-sales. This is referred to as the front office development path of e-commerce since it involves dealing on-line with final customers. Integration of closed business networks, involving suppliers and distribution networks, is defined as the back office development path. The next step is integration of applications and exploiting processes synergies – the all-round e-commerce model.

The transparency of the internet should be a driving force for changing business strategies and attitudes and yet it will take a quantum leap in construction business culture to disclose sensitive information to the supply chain. It has been suggested by a leading construction industry dot.com, that the European construction industry could save up to € 175 billion per annum on building costs and reduce completion time by up to 15 per cent through the widespread adoption of e-construction technologies.

When the lean thinking initiative was introduced into construction, it was the car industry that provided the role model (see Chapters 1 and 3). Now that some sections of the construction industry are seriously talking about e-commerce, it can once again look to the motor industry for a lead. In America, the three major domestic motor manufacturing firms have been dealing with suppliers via a single e-commerce site for a number of years – an initiative that has resulted in reported savings of over £600 million a year. Similar initiatives are also to be found in the retail and agriculture sectors, but perhaps in the rush to establish the first truly

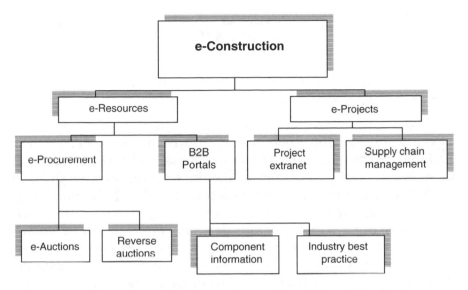

Figure 5.4 The shape of e-construction (adapted from Winch).

successful construction-based e-portal the UK players ignored some basic business rules.

Several years after the advent of e-construction, the current shape can be considered as shown in Figure 5.4.

e-Resources

B2B portals

B2B portals combine a number of easily accessible e-markets and include the following resources:

- industry best practice – for example, www.constructingexcellence. org.uk
- material and component information. www.safelecsupply.com

e-Portals can be broken down into categories, based on who is trading with whom. Latterly sectors such as construction, finance, etc., have coined their own terms, e.g. e-construction, e-finance, to stake their own unique claim in the electronic marketplace. Although some sources claim that there can be up to nine classifications of e-commerce, most people agree that there are only three and of these only two, business to business (B2B) and business to consumer

(B2C), have seen strong growth, the other sector being consumer to consumer (C2C) or person to person (P2P).

Business to business (B2B) or business to administration (B2A), for example on-line data exchange

The extent by which B2B has been adopted by business depends on the sector and the size of the organisation. This category of e-business utilises the internet and extranets and it is forecast that spending in B2B is expected to dominate internet growth until 2010. There are basically two different types of B2B companies, horizontal and vertical. As illustrated in Figure 5.5 vertical B2B companies work within an industry and typically make money from advertising on specialised sector specific sites or from transaction fees from the e-commerce that they may host, for example, BuildOnline.com. Horizontal B2B companies are a completely different breed and operate at different levels across numerous different verticals. Whether it's enabling companies to electronically procure goods, helping to make manufacturing processes run more efficiently, or empowering sales forces with critical information, most horizontal companies make their money by selling software and related services, for example, www.tendersontheweb.com

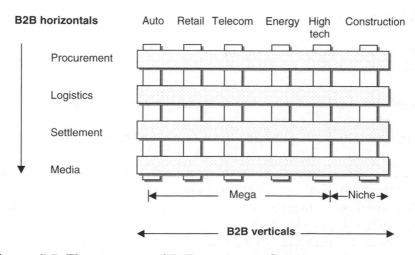

Figure 5.5 The structure of B2B commerce (Source: PricewaterhouseCoopers).

Business to consumer (B2C) commerce

Banking on-line is predicted to rise to include nearly 24 million users by 2007. The financial sector has been an enthusiastic adopter of B2C as bank transaction costs can be reduced by up to 2000 per cent with internet-based services. The travel and accommodation sector, entertainment, shopping on-line including, books, CDs and videos, etc., have followed suit. The motive is simple, added value. It has been estimated that to issue an airline ticket manually costs £5 as compared with 52p via the internet. This sector mainly utilises the internet, but despite the attractions for companies, B2C is considered a highly unstable sector employing, in some cases, questionable accounting practices.

The development of B2B commerce has been rapid compared with other technological innovations. Figure 5.6 illustrates the development of B2B commerce. After the establishment of closed EDI networks, the second phase of e-commerce saw the emergence of one-to-one selling from websites and a few early adopters began pushing their websites as a primary sales channel, e.g. Cisco and Dell. Most of the initial websites were not able to process orders or to supply order tracking. The current phase of development is represented by the

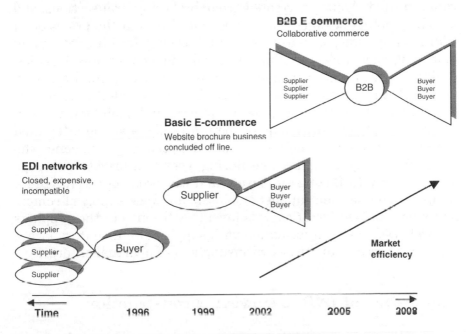

Figure 5.6 The development of B2B commerce (Source: Morgan Stanley Dean Witter internet research).

butterfly model and involves the rise of third party sites that bring together trading partners into a common community. Therefore, in this situation buyers and sellers start to regularly visit the site and move to the collaboration phase, which creates the opportunity to serve a large percentage of those interests. Within a few years, B2B has developed from a high-cost, rigid system with low transparency to a low-cost, highly flexible and fully transparent system.

The development of e-commerce towards e-business, the more comprehensive definition which is now preferred, reveals the move from the enterprise centric vision to the multi-enterprise network, or virtual enterprise and the move towards the exploitation of the ICT and the potential of the internet from cost reduction oriented electric commerce to the collaborative-commerce vision, a move that requires a major step change in business culture. The challenge for construction and the surveying professions is undoubtedly greater. Dell, for example, has the considerable advantage of already being in the internet-related product field, where clients and suppliers are already technologically sophisticated, but nevertheless, real value added benefits are available for the quantity surveyor. The key point which is now currently acknowledged is that with the usage of ICTs and the internet in business, not only costs can be reduced, but also value created. Value is created from brokering transactions and matching orders between companies, but also from the provision of additional services, such as professional services for integrating and managing companies, including legal and financial services, logistics, project and contract management as well as background services, like market intelligence. For example, Dell has created virtual integration with both their upstream partners and their downstream clients so that the entire supply chain acts like a single integrated company. Dell builds computers to order; typically someone who works for a large company like Boeing, goes to a private web page available only to Boeing employees and can order and configure a computer on-line. Dell suppliers maintain a 2-week supply of components near to Dell factories: this inventory belongs to the suppliers not Dell. Dell shares information with suppliers on inventory levels, sales and forecasts and work with suppliers as a virtual enterprise.

Application of B2B to surveying/construction

One of the most discussed topics in electric commerce is business models, because just like conventional business, e-business needs to make a profit and as many dot.com companies have found to their cost, a key component in this process is a robust business model. Today this is more important than ever as the cold wind of reality hits e-commerce. What is needed now is a proven business case; funders want to see profits before they'll hand out any more cash. For when a potential investor in an e-commerce project asks the question 'what is your business model?', what they are really trying to discover is where the business is going to make money and why are people going to pay for using that particular service? A business model should give product/service, information, income generation flows and together with the marketing strategies, enable the commercial viability of the project to be assessed. An e-business model makes it possible to answer the questions such as how is competitive advantage being built and what is the positioning and the marketing mix? In theory very many new business models can be conceived; however, in practice a limited number only are being realised in electronic commerce. Figure 5.2 illustrates 11 business models. They are mapped along two dimensions to indicate:

1. The degree of innovation, which ranges from (in the bottom left-hand corner) essentially an electronic way of carrying out traditional business to value chain integration (in the top right-hand corner), a process that cannot be done at all in a traditional form, as it is critically dependent on information technology to let information flow across networks
2. The extent to which functions are integrated.

Other models which have applications in quantity surveying/construction follow.

Application service providers (ASP)

While the ASP model is new and no settled definition is possible, it can be regarded as a relationship whereby an IT service company, such as Microsoft or Cable and Wireless, manages and delivers applications and/or computer systems to business users remotely, via the internet. This model is to be found in several applications used by quantity surveyors.

e-Shop – seeking demand

This can be the web marketing of a company to promote, in the first instance, the services that it provides. Seen as a low-cost route to global presence, it is now almost obligatory for quantity survey-ing practices to have their own website. Its function is primarily promotion, although it has to be sought by prospective clients.

e-Mall – industry sector marketplace

An electronic mall in its basic form is a collection of websites usu-ally under a common umbrella, for example, www.propertymall. com. The e-mall operator may not have an interest in the individual sector and income is generated for fees paid by the hosted websites which are usually composed of an initial set up fee plus an annual fee. The cost of this sort of platform package can be comparatively modest, depending on the size and the complexity of the website. In addition some marketplace portals permit related organisations (say a quantity surveying practice) to list limited company details for no charge, either permanently, or as an introductory, limited time offer. Construction has large volumes of buyers and sellers, plus geographically dispersed sites. Remote buyers and sellers may be brought together, even when based in different countries. An on-line marketplace cuts the cost of communication, reduces communication errors and speeds up the entire procurement process.

By 2001 it was stated that B2B marketplaces in the construction sector had failed despite over £100 million invested in internet services aimed at the UK construction industry and the existence of more than 80 B2B construction-related sites resulting in too many sites vying for attention. There have been survivors and suc-cess stories, namely BuildOnline.com and eu-supply.com, although both are strictly not simply B2B portals and have moved away from the transactions market to the provision of collaboration platforms. So instead of open market places it seems likely that the construc-tion industry will gravitate towards more private trading hubs. Increasing consolidation is expected to see each specialist area (col-laboration, procurement) dominated by two or three sites.

Perhaps also, a note of realism has been sounded, as contractors now seem to understand that their core business is not suited to running these kinds of ventures and as a result they are turning, as will be discussed later, to specialist providers to supply the technology and organisation for them. In addition, there has been an awareness that companies cannot provide every kind of service; consequently, the future for e-commerce seems to lie, at least within the property sector, with niche market provision. One fact that all sectors of e-commerce are sure about is, that the more fragmented the market, the more efficiency benefits e-commerce ventures can bring, by uniting the disparate elements of the supply chain.

Little wonder, therefore, that amidst the pressure to do business electronically, the process of choosing the right solution can be somewhat confusing, leading to decision inertia among some sections of the professions and industry. However, despite the claims and counterclaims, there is one undeniable fact about electronic commerce and that is the revenues it generates are truly immense and predicted to rise. The question for many construction related services, including surveying, is 'will e-commerce transform business practice from "bricks and mortar" to "clicks and mortar"?' For the surveying practice thinking of launching into e-commerce, it is essential to view the venture holistically – e-commerce is not just the introduction of cutting edge technology, but the integration of technology into existing business plans to introduce new working practices. When drawing up a business plan that includes an e-commerce application, one of the fundamental questions for a surveying practice must be, 'what capital and long-term costs are involved and what return can be expected?' When questioned about the perceived obstacles to adopting e-commerce, organisations replied as follows:

1. *Regulation*. As reviewed at the end of the section under E-commerce law, the regulation and security aspects of a virtual trading environment cause many enterprises concern. In a recent survey carried out among European SMEs by eMarketer Inc., 47 per cent of companies surveyed cited lack of legal guarantees and trust for on-line transactions as a major issue.
2. *Technology and standards*. Which of the alternative standards to adopt, which technology to invest in and which of the systems will be proof against costly upgrades in a few years time?
3. *Costs*. Levels of investment and involvement in e-commerce can vary considerably according to the size and nature of the organisation: from level 1, the entry level, limited to the simple use of electronic commerce requiring minimal investment in hardware, software and human resources, a perfect platform for the small practice through to level three, the virtual organisation

requiring a major investment in human resources as well as industry-wide culture and process change. As illustrated in Table 5.2, the costs associated with entry level to e-commerce are minimal, probably less than a season ticket to watch Arsenal or Manchester United and therefore the perception that involvement attracts a high cost is a false one. Nevertheless, the questions that should be addressed at the outset are:

1. At what level, both in terms of cost and commitment, should an organisation enter e-commerce?
2. Which business model should be adopted?

e–Resources: e-Procurement

Electronic tendering enables the traditional process not only to be made more efficient but also to add significant value. It can provide

Table 5.2 Levels of e-business

Level	1	2	3
Capability	Electronic mail and world wide web	Information management (electronic procurement)	Virtual organisation e.g. Dell, Cisco, Amazon
What is needed?	Personal computer Internet access Software	Process redesign	Business redesign Culture change
Cost	Minimal investment Cost of running parallel systems	Change management	Change management
Benefit	Easy access to information Less information processing	Competitive advantage Streamlined business processes Alignment with supply chain partners	International competitiveness Best practice operations Robust relationships
Who benefits?	SMEs	Public sector Industry	Public sector Industry All members of the supply chain

a transparent and paperless process allowing offers to be more easily compared according to specific criteria. More importantly, by using the internet, tendering opportunities become available to a global market.

e-Procurement is the use of electronic tools and systems to increase efficiency and reduce costs during each stage of the procurement process. Of all the resources referred to in Figure 5.4, e-procurement is the one that intersects the most with other typologies often in a complex way. Since autumn 2002, there have been significant developments of e-procurement: legislative changes have encouraged greater use throughout the EU and new techniques such as electronic reverse auctions have been introduced, not it has to be said, without controversy. In addition, the UK Government has launched a drive for greater public sector efficiency following HM Treasury's publication of the Gershon Efficiency Review: *Releasing Resources to the Frontline* in July 2004, and e-procurement is seen to be at the heart of this initiative.

The stated prime objective of electronic tendering systems is to provide central government, as well as the private sector, with a system and service that replaces the traditional paper tendering exercise with a web-enabled system that delivers additional functionality and increased benefits to all parties involved with the tendering exercise. The perceived benefits of electronic procurement are:

- Efficient and effective electronic interfaces between suppliers and civil central governments, departments and agencies, leading to cost reductions and time saving on both sides.
- Quick and accurate pre-qualification and evaluation, which enables automatic rejection of tenders that fail to meet stipulated 'must have criteria'.
- A reduced paper trail on tendering exercises, saving costs on both sides and improving audit.
- Increased compliance with EU Procurement Directives, and best practice procurement with the introduction of a less fragmented procurement process.
- A clear audit trail, demonstrating integrity.
- The provision of quality assurance information – e.g. the number of tenders issued, response rates and times.
- The opportunity to gain advantage from any future changes to the EU Procurement Directives.
- Quick and accurate evaluation of tenders.

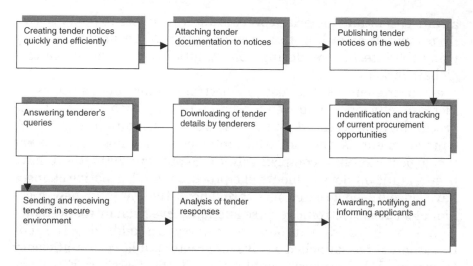

Figure 5.7 e-Procurement.

- The opportunity to respond to any questions or points of clarification during the tendering period.
- Reduction in the receipt, recording and distribution of tender submissions.
- Twenty-four hour access.

Figure 5.7 illustrates the possible applications of e-procurement to projects that are covered by the EU public procurement directives. Benefits of electronic tendering and procurement of goods and services are said to be wider choice of suppliers leading to lower cost, better quality, improved delivery, reduced cost of procurement (e.g. tendering specifications are downloaded by suppliers rather than by snail mail). Electronic negotiation and contracting and possibly collaborative work in specification can further enhance time and cost saving, and convenience. For suppliers, the benefits are more tendering opportunities, possibly on a global scale, lower cost of submitting a tender and possibly tendering in areas which may be better suited for smaller enterprises or collaborative tendering. Lower costs can be achieved through increased greater efficiency and in some sectors the time may not be too far distant when the majority of procurement is done this way. However, a survey carried out by e-Business Watch in 2002 over 6000 organisations found that nearly 60 per cent of those surveyed perceived that the lack of face-to-face interaction was a barrier to e-procurement, while on-line security was still a major concern.

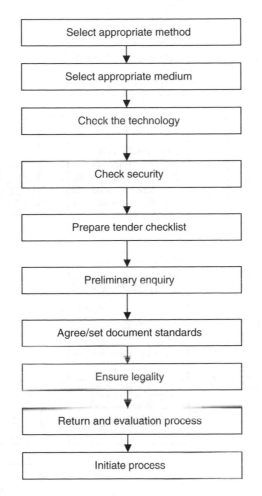

Figure 5.8 e-Procurement process (adapted from RICS).

In autumn 2005 the RICS produced a guidance note on e-tendering in response to the growth in the preparation of tender documents in electronic format. Figure 5.8 sets out their recommended approach to the e-procurement process, while Figure 5.9 maps the way by which contract documentation can be organised for the e-tendering process.

e-Auctions

An on-line auction is an internet based activity, which is used to negotiate prices for buying or selling direct materials, capital or services.

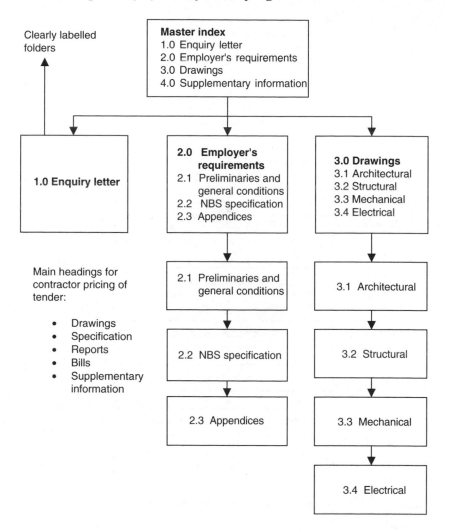

Figure 5.9 Organisation of contract information for e-procurement (Source: RICS).

On-line auctions can be used to sell: these are called forward (or seller) auctions and closely resemble the activity on websites such as eBay; the highest bidder wins. Now some companies are starting to use reverse auctions where purchasers seek market pricing inviting suppliers to compete for business on an on-line event see Figure 5.10. Auctions can either be private/closed where there are typically few bidders who have no visibility of each other's bids, or open, where a greater number of participants are invited. In this case participants have visibility of either their rank or the bidding itself.

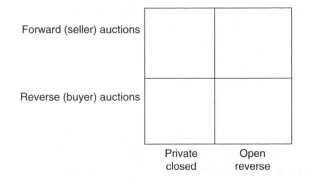

Figure 5.10 e-Auctions.

When used, the technique can replace the conventional methods of calling for sealed paper tenders or face to face negotiations.

On-line auctions are said to offer an electronic implementation of the bidding mechanism used by traditional auctions, and systems may incorporate integration of the bidding process together with contracting and payment. The sources of income for the auction provider are from selling the technology platform, transaction fees and advertising. Benefits for suppliers and buyers are increased efficiency and time savings, no need for physical transport until the deal has been established, and global sourcing.

There have been some strong objections to e-auctions and in particular reverse e-auctions from many sections of the construction industry. In the public sector, the OGC has received representations from trade associations and other bodies. Sections of the industry have seen e-auctions as a return to lowest price purchasing, threatening already low margins. The industry also perceives that e-auctions as challenging the principles of many government-led initiatives (see Best Practice portals) such as an integrated supply chain approach to construction procurement based on optimum whole-life value. Amongst quantity surveyors, the perception of reverse e-auctions is very negative with 90 per cent feeling that they reduce quality and adversely affect partnering relationships.

Reverse auctions

The reverse e-auction event is conducted on-line with pre-qualified suppliers being invited to compete on predetermined and published award criteria. A reverse auction can be on any combination of criteria, normally converted to a 'price equivalent'. Bidders are able to introduce new or improved values to their bids in a visible and competitive environment. The duration of the event will be defined before the start of the reverse auction commences. There will be a starting value that suppliers will bid against until the competition closes.

Three characteristics that need to be present to have a successful reverse auction are as follows:

- The purchase must be clearly defined
- The market must be well contested
- The existing supply base must be well known.

These three factors are interdependent and together form the basis for an auction that delivers final prices as close as possible to the true current market price. For the buyer and the supplier, a clearly defined scope of work is essential. Without this it becomes very difficult to bid accurately for the work. Contestation, that is three or more suppliers within the market willing to bid for the work, is another prerequisite. Without this, there is no incentive for suppliers to reconsider their proposals. Finally the client's knowledge of the supply base ensures that the most suitable suppliers participate in the events. In order to move away from a system where cost is the only selection criteria, it is possible to organise bidders to submit their proposals on other matters, such as safety or technical ability before the commencement of the reverse auction. These proposals can then be evaluated beforehand and the resulting scores built into the auction tool. Therefore, when a supplier enters their price, the application already had the information needed to complete the evaluation process, the process is known as a transformation reverse auction (TRA) and it is thought to more accurately reflect the prevailing market dynamics.

Digital certificates and electronic tendering

e-Commerce growth rates are critically dependent upon both consumers and businesses having trust in undertaking transactions on-line. As mentioned earlier, trust, or to be more accurate the lack of trust in internet commerce, is a major concern for most organisations. Widespread adoption of e-commerce depends on users having trust and confidence in the whole activity. Businesses have increasingly looked for solutions that provide trust between parties. Digital certificates are critical tools that guarantee the user's identity. Typically a digital certificate is only issued after completion of various checks, designed to verify the user's identity. Typical verification sources include passports and electoral rolls for individuals, and Companies House and alternative corporate registration documents for organisations. The strength of the identity checks made before the certificate is issued are such that substantial liability cover can be offered. So if digital certificates are such a great idea, why are they not so widely available? This is primarily down to lack of availability of relevant applications that drive their adoption. However, one application where they are being rapidly introduced is in the area of electronic procurement. Digital certificates are used not only to confirm the identity of the sender but also to digitally sign and encrypt documents so that they can be sent in a highly secure manner. Both purchasers and suppliers use smart cards to access the system.

The purchaser's smart card enables only authorised people to create and publish the tender documentation to the system, and it also enables them to open the tender responses once the documents are released after the closing date. The supplier's smart card enables them to retrieve the full tender documentation and to submit their responses securely. Suppliers can have total confidence that their bids are completely secure, signed and sealed, until the tender closing date.

A key benefit associated with the use of smart cards is that it provides an irrefutable audit trail underscoring the transparent nature of the process. Every access to the system, and every document movement between purchasers and suppliers is tracked with the automatic generation of receipts at critical points in the process. In addition to the downloading and uploading of documents, the smart card is used to manage access to on-line supplier forums. Only those directly involved in a specific tender can post and view replies clarifying the tender. Electronic tendering using smart card technology offers substantial benefits for both purchasers and suppliers. For purchasers, it means tenders can be downloaded securely at the touch of a button enabling a shorter more efficient tendering process. Other benefits include electronic publication of opportunities to an increased supplier base, supplier selection and qualification, improved supplier communication, and faster availability of tender documentation. Substantial cost and time savings can be obtained by the elimination of the traditional paper chase enabling greater concentration on areas where real value can be delivered such as tender evaluation. In addition there is an irrefutable audit trail of the entire tendering process from submission to award.

Suppliers are not only able to receive relevant invitations to tender as soon as they are uploaded by the purchaser, but they can send back their proposals securely and confidentially, safe in the knowledge that their bids can only be viewed at the tender closing date. They can also make changes, updating proposals right up to the closing date.

How smart cards work

Step 1

- The purchaser prepares the tender documentation using the proprietary software. Typically this will include a notice (summary of the tender) plus associated documentation.
- The purchaser uses the smart card to log on to the system. Their digital signature is checked to ensure that it is valid and that they have the rights to publish tenders on the system.
- The tender documents are then sealed, electronically signed, encrypted and sent to the secure server. Arrival at the server is time stamped, and e-mail notification of receipt is sent back to the purchaser.

Step 2

- At the secure server the notice is published on to the public internet site. The full document set is published to the secure server. (Only suppliers with a valid smart card can retrieve the full documentation.)
- Those suppliers who have set up a tender alert on the system and whose section criteria matches that of the notice are automatically notified by e-mail that there is a new tender opportunity.

Step 3

- Access to the full tender documentation is available to suppliers using their smart card. Before the documents are downloaded the digital certificate on the smart card is validated.

Step 4

- The supplier prepares their tender response attaching the appropriate documents. Different document types can be used, word processing, CAD, spreadsheet, etc.

- The supplier logs on to the system, the validity of their digital certificate is checked. The package of documents is sealed, electronically signed, encrypted and sent across the internet to the secure server. Arrival at the server is time stamped, and e-mail notification of receipt sent back to the supplier.
- The responses are held securely until the closing date and these cannot be accessed.

Step 5
- At the closing date, the purchaser, using their smart card, can retrieve the tender responses from the suppliers for analysis and award. Again, the digital certificate is validated.

Digital certificate trust and security benefits

- Provides controlled access to the e-tendering system to those who have a valid certificate
- Confidentiality – tender documents cannot be read by an unauthorised party
- Non-repudiation – neither party can deny having sent or having received tender documentation
- Integrity – no alterations can be made to the documentation
- Authentication of identity. To register, purchasers and suppliers have to provide detailed information to confirm their identity.

Key purchaser benefits

- Electronic publication of opportunities to an increased supplier base
- Tender documentation can be worked upon right until the last minute. Improved preparation time will increase the likelihood of quality responses
- Invitations to tender at the touch of a button. These tenders will be accessible to all eligible suppliers to upload as soon as the purchaser agrees the issue date and time
- Suppliers' tender responses are automatically received at the tender closing date
- Reduction in paper chase and process
- Enables concentration on value adding activities such as tender analysis rather than administrative tasks such as copying, collating, binding and distribution of tender information
- Provides a complete irrefutable audit trail of all document movements to and from suppliers. An instant record of when documents are sent and received with associated receipts

- Improved management information about tendering activity and process
- Faster and easier communication with suppliers
- An integrated rather than a fragmented tendering process
- Faster pre-qualification of suppliers, ranked on predetermined criteria
- Shorter tendering timescales
- Costs savings in a more efficient tendering process plus better value as a result of increased competition from the supplier base.

Key supplier benefits

- Suppliers can register their profile so that they are notified of invitations to tender specifically relevant to their business
- Comprehensive search of all tender opportunities published. This reduces costs and minimises missed opportunities
- Suppliers can make changes, update proposals and provide further information right up to the tender closing date, ensuring the very best proposal possible is submitted
- Responses can be sent back at the touch of a button, increasing the amount of time spent on proposals, rather than the administrative process of printing, collating and distributing binding documents
- Reduction in the paper chase, with no need to provide multiple copies of tender documents
- Suppliers will be able to concentrate their efforts on value adding activities that could result in winning business, rather than copying, collating and binding tender documents
- Easier communication with the purchaser through an electronic forum that facilitates amendments, commonly asked questions and answers
- Suppliers will have an instant record of when all documents are sent and received. Confirmation of receipt of documents will also be issued. This guarantees a record and assurance of document transmission.

Cost savings

Market studies have shown that savings with electronic tendering systems arise in two main areas:

1. Improvement in the efficiency of the tendering process
2. Better bids leading to reduced procurement costs.

Merx, a Canadian e-tendering notification system (not smart card based) has reported savings of around 15 per cent, just on the improvements on tendering publication and notification. Savings can be achieved in the following areas:

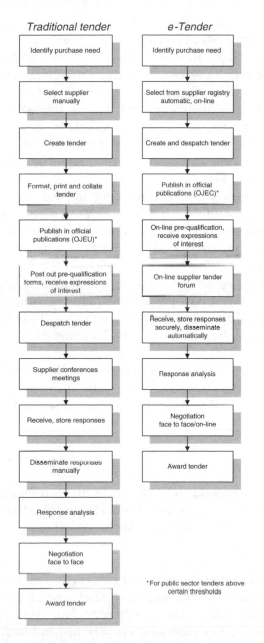

Figure 5.11 e-Tendering versus traditional tendering.

- *Time and cost of tendering.* A shorter tender preparation cycle, reduction in paperwork and improved communication (no couriers) can give savings of around 15–20 per cent.
- *Improved quality of the tender issued.* Less time on the administrative process of printing and issuing tenders means that more time can be spent on the quality of the tender being issued. A better tender reduces the need for management time for supplier meetings to clarify bids and reduces phone calls to check on tender progress. This can provide potential savings in management time of up to 20 per cent.
- *Faster analysis.* Electronic analysis can perform pre-screening; scoring and rating of tender responses to filter out those that are uncompetitive. Management time can be focused on improved evaluation and more detailed negotiation to get the best value from suppliers.
- *Improved competition.* e-Tendering increases the pool of supplier responses since it removes one of the main barriers to participation, namely finding out about the tender. The costs savings through increased competition is estimated at around 15 per cent.
- *Better quality of tender response.* Assistance can be provided through the use of wizards and proposal automation tools to improve the quality of the supplier response. This leads to savings in purchaser's management time in analysis.

The Office of Government Commerce and Barclays Business have developed the Visa purchase card to be used by government departments for low value, high volume procurement.

e-Projects

e-Commerce can be said to be the ultimate supply chain communication tool as it permits real-time communication between the members. e-Commerce can include the use of some or all of the following technologies:

- *The internet.* The international network. The main advantages of the internet for the quantity surveyor include availability, low cost and easy access, whereas the main disadvantages particularly for business users centre around lack of control, reliability and security aspects that are now being addressed and will be discussed later in this chapter. For the surveyor, internet applications include procurement, marketing, e-mail and data transfer. A significant number of hard pressed UK quantity surveying practices unable to source staff in the UK, transfer project drawings in CAD format, via the internet to practices worldwide for the measurement and preparation of bills of quantities. The completed bills plus drawings are then e-mailed back to the UK, permitting 24-hour virtual working.

- *An intranet*. An internal network that publishes information available to staff within a single company, not the world. Compared with the internet, intranet sites are much faster to access and offer great savings in set up, training, management and administration. An intranet is a very cost-effective way of centralising information sources and company data, such as phone lists, project numbers, drawing registers, quality procedures as well as allowing the use of internal e-mail. Intranets use the same technologies as the internet, but are not open to public access.

- *An extranet*. An extranet is a wide area intranet that spans an organisation's boundaries, electronically linking geographically distributed customers, suppliers and partners in a controlled manner. It is a closed electronic commerce community, extending a company's intranet to outside the corporation. It enables the organisation to take advantage of existing methods of electronic transaction, such as electronic data exchange (EDI) a system to facilitate the transmission of large volumes of highly structured data. Project extranets have been described as the first wave of the e-commerce revolution for the UK construction industry and applications that utilise extranets include project management, for example the construction of Hong Kong's new airport. For some, electronic data interchange promised the ability to exchange data efficiently between trading partners, as had been the case in the motor industry and food retailers for a number of years. The major disadvantages of EDI are the high cost, as operators have to trade through value added networks (VAN), as well as standards problems, i.e. not all EDI systems are compatible. More importantly, the point to point contact of EDI provides no community of market transparency. These problems are increasingly being addressed by the reduced costs of internet applications that are able to deliver flexibility, reduced training and low set-up costs. An example of the limitations of EDI is the experiences of H&R Johnson Tiles, the largest manufacturer of tiles in the UK. This company has carried out electronic trading for years with its larger customers, using EDI; however, they have found it impossible to extend EDI to smaller customers without the critical mass of transactions to drive the necessary investment; a separate website had to be launched using an extranet to cater for its top 20 smaller customers. This is not to say that batch-mode EDI transactions will not survive and prosper, as the system is extremely efficient and has been predicted to have an established place in the large scale exchange of data. Extranets/EDI are perhaps the most diffused type of e-commerce and should be better defined as e-business rather than e-commerce.

- *Collaboration platforms*. These provide a set of tools and the information environment for collaboration between organisations. This can focus on specific functions, such as collaborative design and engineering or providing project support with a virtual team of consultants. Business opportunities are in managing the platform and selling the specialist tools. A value chain (Figure 5.12) has been defined as a model which describes a sequences of value-adding activities of a single organisation, connecting an organisation's supply side with its demand side and includes supporting activities. Information technology is applicable at all points of the value chain.

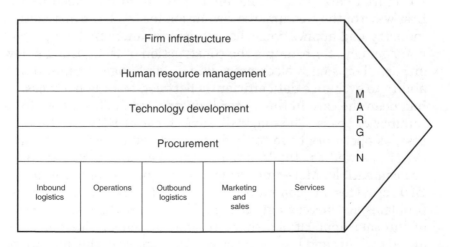

Figure 5.12 Porter's value chain.

Primary activities

Inbound logistics; just-in-time activities;
Operations; process control;
Outbound logistics; on-line link to customers;
Marketing and sales; laptops for direct sales;
Service; electronic dispatch of technical supports

Support activities

Firm infrastructure; e-mail;
Human resource management; On-line personnel base;
Technological development; CAD/CAM;
Procurement; on-line access to suppliers' inventory.

Value chain integration can use internet technology to improve communication and collaboration between all parties within a supply chain.

- *Value chain service provider.* These organisations specialise in a particular and specific function of the value chain – for example, electronic payment or logistics. A fee or percentage based scheme is the basis for revenue generation.
- *Value chain integrators.* These focus on integrating multiple steps in the value chain, with the potential to exploit the information flow between these steps as further added value.
- *Virtual communities.* Perhaps the most famous of virtual communities is the ubiquitous Amazon.com.

The system illustrated in Figure 5.13, is a level 2 application (see Table 5.2) and allows savings and efficiency gains to be made through such measures as reducing the amounts of abortive or repeat work carried out by the contractor and allowing manufacturers to employ just-in-time production techniques. Any changes that are made to the project information, for example alteration to the specification, are immediately communicated to all parties. E-commerce allows true transparency across the supply chain permitting the sharing of information between members in real time.

A further example is given in Figure 5.14 where the traditional lines of communication between the various supply chain members is illustrated. Without the presence of a hub to link and co-ordinate the various parties, decisions can and frequently are, taken in

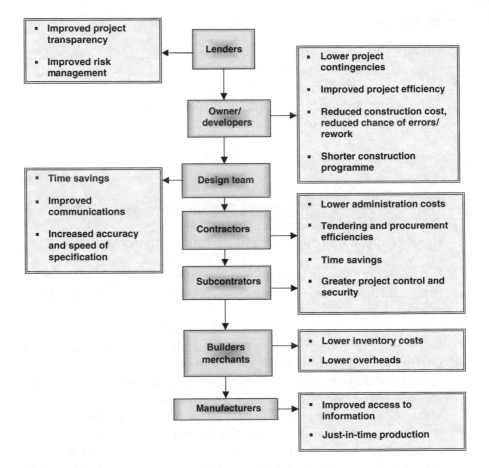

Figure 5.13 e-Commerce and the supply chain (Source: e-Business for the Construction Industry – BuildOnline.com).

isolation, without regard to the knock on effects of cost, delays to the programme, impact on other suppliers, etc.

However, by contrast the introduction of a collaborative hub, for example, the packages offered by the Building Information Warehouse http//:www.biw.co.uk permits decision-making to be taken in the full light of knowledge about the possible implications of proposed changes. In addition, it also permits specialist subcontractors or suppliers to contribute their expertise to the design and management process, thereby liberating the potential contained within the supply chain (see Figure 5.15).

In a recent case study, the quantity surveying practice, Gleeds, calculated the drawing production cost savings arising from the use of an electronic data management collaboration system used

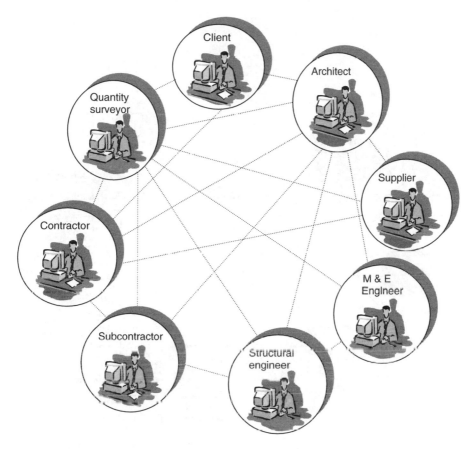

Figure 5.14 Traditional supply chain communication.

on a £5 million, 30 week retail construction project were as in Table 5.3 (p. 201).

The quantity surveyor and e-commerce

For the purposes of this review, e-business has been broken down into the following:

1. How the surveyor can participate in e-commerce
2. Guidelines for the integration and adoption of e-commerce practice.

The quantity surveying profession could never be accused of being Luddites and information technology already plays a large part in

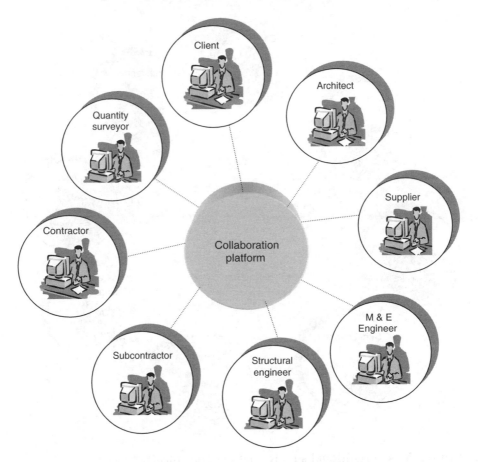

Figure 5.15 Collaboration platform communications

the development process. However the DETR report IT Usage in the Construction Team, found that although the majority of information in the construction industry is created using IT, most is distributed in paper form. The report found that:

- 79 per cent of specifications are produced electronically against which 91 per cent are distributed on paper.
- 73 per cent of general communications across the design team are produced electronically, yet 85 per cent are distributed on paper.
- Only 5 per cent of contractors' tenders are received in electronic format.

It's little wonder that the European construction industry has been estimated to spend £300 million a year alone on couriers! In addition BuildOnline claim that lost paperwork and lack of

Table 5.3 Example of savings using a management collaboration system

Printing cost for project drawings	£46 112
Postage for drawings	£1584
Copying costs for project specification	£10 215
Postage costs for specification	£219
Total	**£58 130**

Source: CITE.

communication adds 20–30 per cent to construction costs across the board. Tables 5.2 (p. 182), 5.4 and 5.5 have been included in order to give the surveyor some indication of the potential for the integration of e-commerce into day-to-day practice. Three levels have been proposed based upon commitment, ranging from level one, requiring minimal investment, but nevertheless still at a level capable of producing tangible benefits and savings to both clients and business, through to level three, the virtual organisation requiring total commitment and a high level of investment.

Entering the world of e-commerce should be a considered business decision, not a knee jerk reaction to the sudden availability of new technology. An existing bricks and mortar company's goals are different from those of an idea company. Likely to be already profitable, a bricks and mortar company will most often turn to the internet to expand its markets, meet customers' needs and improve operating efficiencies in ways that are usually incidental to an existing business – not to reinvent its business. The first step is to identify the business goals that can be served by using the internet. Answering

Table 5.4 e-Business level 1

Applications	*Benefits*
1. E-mail	1. Real-time communication
2. Marketing	2. Access to new markets on similar footing to larger organisations
3. Participation in electronic procurement/auctions	3. Clients may increasingly insist on electronic procurement
4. Access to databases for example: TED Tenders direct Barbourexpert Constructionline	4. Information on competitors/market opportunities Sourcing using web pages

Table 5.5 e-Business level 2

Applications	Benefits
1. Fully electronic procurement including: • Pre-qualification • Comparing bids • Evaluation • Contract award • Archive information on bidders including KPI rating • On-line auctions	1. By far the biggest cost savings come from the reduced costs of creating and disseminating tenders; evaluation of bid responses, creating purchase orders and tracking progress Access to benchmarked records on contractors and supply chain
2. Fully electronic tendering including: • Tender submissions • Submission of bids • On-line estimating • Exchange of information • Payment	2. Reduce wastage, increase profit margins Reduce errors
3. Project management • Dissemination of information	3. Effective supply chain management Real-time communication Integration of the supply chain
4. Enables virtual design teams • Collaboration with practices worldwide	4. Contracted out staff could be located anywhere on the globe

this question depends in part, upon the nature of the business. For example, How large is the company? What sort of distribution chain does it utilise? Is its customer base static or could it be expanded? Some examples of potential e-commerce applications to surveying practices over levels one to three are shown in Tables 5.4–5.6.

Even at the entry level considerable market advantage can be gained through the use of e-mail and interactive websites.

Table 5.6 e-Business level 3

Applications	Benefits
1. e-Construction packages including project collaboration packages allowing construction professionals to access and amend project information	1. e-Construction packages Alter and amend project documentation at minimal cost Ensure that everyone works on the most up to date information A repository for comments Provide accurate audit trail Compress the construction programme Few change orders
2. Industry and supply side marketplaces	2. Real-time information of stock levels/delivery, etc.
3. Contractor consortia	3. Pooling and sharing of information
4. Niche markets	4. Dispute resolution

The choice of partner can be crucial in maximising opportunities. For example by working with existing operators, it is possible to start benefiting immediately.

A constant theme throughout this book has been client criticisms of the UK construction industry and its fragmented structure. Clients, professionals as well as the entire downstream supply chain try to cope with the challenge of operating in a highly frag- mented industry, where the top five contractors own less than 10 per cent of the marketplace. The fragmentation has over the years contributed to poor profitability and cash flow even for the major players which have in turn prevented investment in new technolo- gies as noted by Latham and Egan. Add to this the constant de- mand by clients for added value and the case for adopting e-com- merce solutions and practices seems irresistible. As demonstrated in Figure 5.13, the biggest benefits of e-commerce are likely to come from the integrated supply chain, where information is freely available between clients, contractors and suppliers.

Legal aspects of e-commerce

Beyond the basic business considerations, a host of legal issues faces an industry or profession preparing to go on-line. This review of the implications of e-commerce for surveying practice cannot be concluded without mentioning some of the practical legal considerations that must be taken into account.

An aspect of e-trading that has caused considerable concern in both B2B and B2C sectors is the regulation of transactions in a global market, where organisations and governments can find themselves dangerously exposed in what has been described as a virtual 'wild west' environment. There is also the view, held by many in the IT sector that the infrastructure is just not robust or sufficiently mature enough to do the job being asked of it as a result of the e-commerce hype.

The intersection of a global medium, like the internet, with systems and legislation designed for the physical, territorial world, poses many problems. Compared with other entities, the internet has developed in a spontaneous and deregulated manner and does not have a central point of authority. In the event of contested claims and possible litigation, the fundamental problem of jurisdiction remains unresolved, as does the security of the systems, although as previously mentioned steps are being taken to establish the integrity of cyber space with the introduction of cryptography and digital signature services. Its technical development has been guided by protocols established through bodies such as the Internet Engineering Task Force, but there has not been a central rule making body that has exercised comprehensive legislative authority over the internet, and there is unlikely to be one. The multi-jurisdictional and multi-functional nature of the internet means that, inevitably, many different interests in many different parts of the world will be concerned with any endeavour to formulate specific policies. Even in the European Union, the suggestion contained in the Commission's directive that e-commerce should be governed by the law of the country where the service provider is established has been questioned by consumer groups who want the local law where the website is accessed to be given priority.

Another crucial area of concern, primarily for governments is e-tax or the tax treatment of on-line transactions. At present the volumes of e-transactions means that the fiscal implications are modest. However, if predictions of growth are to be believed, the question of how and which government is to tax such revenues could be very important. Within the OECD area views diverge. In the USA, the belief is that e-transactions should not be taxed, while in the European Union the view is that VAT should be levied on e-trading. The law in the field of e-commerce is continuously developing and fast moving, as numerous drafts pass into the statute books. The evolution of technology also means that legislation must be updated and requires constant review. Organisations need to be aware of both the current and the prospective impact of legal provisions.

The principal regulatory concerns are focused in four areas: on-line contracting and security, both dealt with previously in the review of encryption and private key services, as well as regulation/jurisdiction and intellectual property protection.

The implementation of electronic keys is dealt with in an EC Framework Directive implemented in July 2001. The UK E-Communications Act 2000 when originally drafted, contained powers to be vested in law enforcement agencies, requiring disclosure of electronic keys to decrypt information where necessary. This provision caused considerable debate as it was seen by many as a major barrier to the promotion of the UK as a favourable environment for e-commerce. It was suggested that the Home Office had hijacked the bill and eventually the government was forced to delete the provisions, although in practice they were only moved to the Regulation of Investigatory Powers Bill, which is now on the statute books. The UK E-Communications Act therefore is now quite simple in that it allows for:

- The introduction of a new approvals regime for providers of cryptography services
- Electronic signatures to be admissible in court
- The updating of existing legislation to allow the use of electronic communications.

In the early days of e-commerce, it is true to say that the legislative framework was lagging behind business practice. How can an organisation be sure that information being transmitted electronically is secure from its competitors?

The major legislative instruments in e-commerce law are:

- The UK Electronic Communications Act 2000
- The Electronic Commerce Directive (00/31/EC) – 2002
- The UNCITRAL Rules – final draft expected 2006.

It does, however, only apply to services supplied by services providers within the EU. Countries outside the EU are covered by UNCITRAL rules. The e-commerce directive establishes rules in the areas including definition of where operators are established, transparency obligations for operators, transparency requirements for commercial communications, conclusion and validity of electronic contracts, liability of internet intermediaries and on-line dispute settlement. Put simply, the directive states that the service providers are subject to the law of the member state in which they are established or where the ISP has its 'center of activities'. B2B contracts are of particular importance as national laws govern the main aspects of contract law and what constitutes an offer or an acceptance varies from country to country. For example, the rules formed under English law on offer and acceptance – at what point is the offeree's acceptance communicated to the offeror. In addition, communication by website is instantaneous, whilst e-mail is not. The possible scenarios include the offeror failing to collect e-mail from the server or failure of the ISP.

European Directive 00/31/EC attempts to clarify the situation by stating that the contract is concluded when the offeree is able to access the offeror's receipt of delivery. Unfortunately this clause does not cover the position, say, of an invitation to tender which is an invitation to treat.

In general, the parties to the contract should agree by what is called private autonomy as to which country's law of contract is to apply.

e-Contracts

Contrary to popular belief, there is no legal reason to prevent a binding legal contract being made by way of e-mail. Provided that there is a clear offer and acceptance of all the critical terms. Therefore, within a quantity surveying practice for example, employees should be reminded of any limits on their authority to conclude contracts on-line and this could extend to adding a standard disclaimer to e-mails to this effect. Recent legislation in the B2C sector, implemented by the UK Government now applies to contracts that are concluded at a distance between say a construction supplier and a consumer. The website of the supplier must now include:

- The identity and postal address
- Characteristics of the relevant goods/services
- The price, delivery costs and taxes
- The existence of the right of withdrawal from the contract.

Failure to comply with these conditions gives the consumer the right to cancel the contract within 7 days.

Conclusion

There can be no doubt that the future for both business in general and for the quantity surveyor in particular, is as part of the digital economy. In the late 1990s, revolutionary new business models were set to destroy old economic values, but only 2 years later the talk of the collapse of the new economy was just as overstated. However, to equate the downturn in the e-economy with the demise of the internet is like the pundits who proclaim the death of the quantity surveyor – very exaggerated! The market is maturing and as has been demonstrated in this chapter, the shift towards the digital economy is unstoppable and there are real benefits that can be brought to a fragmented industry by e-commerce.

Therefore despite all the pitfalls, both technological and legal, that have been outlined in this chapter what are the critical success factors to be considered by an organisation still determined to be part of the digital revolution? First, a little reassurance. It would also appear that e-commerce security concerns may have been overstated. As indicated in Figure 5.16, the latest data indicate that the most widespread information society breaches are computer virus infections. Almost all organisations have been affected by computer viruses. By comparison the number of businesses affected by other security breaches such as unauthorised access to their networks or identity theft are fairly low.

Therefore, assuming that the starting point for a new e-commerce venture is all-round e-commerce (see Table 5.4), the points to consider are as follows (these will vary in line with the level of entry, see Table 5.2):

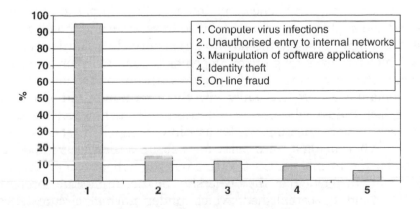

Figure 5.16 Types of security breach (Source: SIBS).

- *Content.* This should be a unique service that exploits the electronic environment and delivers added value to potential users. The presence of a unique and/or innovative product or service that is saleable over the internet, that is, it fits with the demographics of the internet. In addition, the platform must be capable of attracting sufficient clients to generate cash flow to repay start up and running costs. For example, dispute resolution on-line.
- *Community.* There must be the ability to build up a critical mass of customers/business partners for the venture which will translate into sales/cost savings to cover the initial investment. It is the inability to establish a community of clients that has caused so many dot.coms to fail.
- *Commitment.* Clear objectives are vital. Most clearly demonstrated by a defined business case for the e-commerce venture, but at the very least, a clear idea of objectives and the demonstration of strong motivation for using the internet. For example, just-in-time production for building materials – a large volume of curtain walling could be produced only when it is required for incorporation into a project. The manufacturer in return can tap into a global supply base for raw materials.
- *Control.* This is the extent to which e-commerce is integrated with internal business process, enabling the organisation to control all aspects of its business and handle growth and innovation.

Website design

The address of the website is crucial as it will influence the number of potential clients who are able to find it. Although domain name registration is relatively cheap and quick, about £5 per annum for a '.co.uk' and slightly more for a '.com' address, time and money can be lost if a similar name has already been registered by a third party. The English courts have given firm judgements against cyber squatting, which is the practice whereby a person with no connection to a specific domain registers it, usually with the intention of selling it on to the highest bidder who has a connection with it, however except in the case of cyber squatting, the law is still very much, 'first come, first served'. When using a third party for website design, it is important to establish a formal contract setting out the obligations of designer and client. In particular, the ownership of the intellectual property rights should be established, which under normal circumstances should rest with the client. Website content must comply with the Data Protection Act. Five years ago there were five or six major search engines, now the market is dominated by one – Google. The message is clear – a site will not become noticed if it is not ranked by Google. Five or six years ago submission of a website to the search engines was free of charge, today most search engines require up-front payments of several hundreds of pounds to guarantee consideration and ranking. There are a number of companies such as ineedhits.com who will, for a small fee, submit a website to the important search engines on a monthly basis.

Bibliography

Building Centre Trust (2001). *Effective Integration of IT in Construction*, London.

Cartlidge, D. (2004). *Procurement of Built Assets,* Butterworth Heinemann, Oxford.

Chappell, C. and Feindt, S. (1999). *Analysis of E-Commerce Practice in SMEs*, Kite.

Department of Trade and Industry (1999). *Building Confidence in Electronic Commerce – A Consultation Document.*

e Building for Clients (2005). Department of Environment, Trade and the Regions.

OGC (2005). *e-Procurement in Action – A guide to e-Procurement in the Public Sector,* Office of Government Commerce.

Royal Institution of Chartered Surveyors (2005). *E-tendering*, RICS Books.

SIBIS (2003). *Matching up to Information Society*, Statistical Indicators Benchmarking, The Information Society.

Timmers, P. (1999). *Electronic Commerce – Strategies and Models for Business to Business Trading*, Wiley & Sons Ltd.

(2000) Industry e-commerce to hit £67bn by 2004, *Building* 6 October.

Websites

www.ogc.gov.uk/
www.europa.eu.int
www.e-envoy.gov.uk
www.dti.gov.uk
www.buildingcentretrust.org
www.cite.org.uk
www.itcbp.org.uk

6

New technology – opportunity or threat?

Graham Castle MSc FRICS

Introduction

During the preceding chapters, the opportunities for quantity surveyors to develop new expertise in areas that are either available now or emerging have been described. Many of these opportunities depend on the adoption of new technologies. This chapter reviews other industries' experiences of integrating new technology into existing work practices and processes, and concludes with a strategy for successfully integrating new technology into a quantity surveying organisation.

A bottom-up approach

Technology has had a profound impact upon other industries over the last decade, but not on the construction industry; why is this? This chapter sets out to answer this and other important questions, such as how to employ new technology in the workplace, and where to use it.

The adoption of new technology has changed the banking industry beyond all recognition and has led not only to the development of new services but also, more importantly, to new ways of working. Virtual money may not yet be a reality, but the virtual bank is. Bank managers have become an endangered species, and one can make payments and withdrawals 24 hours a day, 365 days a year using an automatic cash machine or the internet. Indeed, it is possible to do all of your banking today without having any face-to-face contact with another human being.

However, the banking industry is not an isolated example. Twenty years ago, the newspaper industry was, if anything, even more conventional in its ways than the construction industry is today. It was only after a long, protracted and sometimes violent industrial dispute that employers were able to introduce new working practices based upon new technologies that enabled digital publishing. The newspaper industry's experience exemplifies the fact that new working practices and the allied new technologies are not always welcomed into the workplace by everyone. In the case of the newspaper industry, the emphasis was on the introduction of new digital technology and the elimination of inefficient historical working practices, combined with the irrelevance of some traditional trades. Another example of the adoption of new technology is the automobile industry, discussed in Chapter 1, where once again the introduction of new working practices such as computer-aided design and manufacture, just-in-time materials deliveries, electronic ordering and bill payment has enabled manufacture on demand. Here the emphasis has been upon the integration of design and manufacture, supply chain management, and the introduction of robotics.

These examples clearly show that in order to gain the full benefits of new technology, working practices have to change. Therefore, it seems likely that the quantity surveyor is going to have to do more than merely become competent with a spreadsheet to take advantage of these new technologies. It is also worth noting that all of these so-called revolutions in other industries occurred before the age of the all pervasive internet!

Historically, quantity surveyors have not been slow to adopt technology. They were amongst the forerunners in the construction industry to adopt personal computers and software such as spreadsheets, bills of quantities production software and word processing. However, one problem associated with the introduction of computers into the workplace was that technology competence tended to decrease in direct proportion to seniority within an organisation. This lack of knowledge of new technologies on the part of senior surveyors, the very people who are identifying the business objectives and managing the firm, led to scenarios where investments in new technology were driven from the bottom up within companies. Typically, a graduate surveyor would identify a task that could be made more efficient or less tedious by the application of a new technology. This led, often after considerable persuasion, to a purchase being made to satisfy the individual user's need. Another typical scenario was that a nearby competitor would acquire some new technology and it was then felt that the firm should do likewise so as not to be left behind. This approach led to islands of computing being established within firms, where new technology was acquired upon a task basis with little or no attention being paid to the overall strategy of integrating new technology into the workplace for the benefit of the business overall. Not surprisingly, many of these investments failed to provide the expected returns, and in some instances they were even counter-productive, leading to a fall in productivity.

Although the examples given at the start of this chapter regarding acquiring new technology are all on a grander scale than that of the typical quantity surveying firm, they are all typified by being championed at the highest levels of management within their respective organisations. This does not mean that senior managers in the banking, newspaper or the automotive industries are more technology literate than senior surveying managers. It does, however, identify that they recognised how new technology could be allied to their business objectives, and then championed that cause through thick and thin, sometimes ruthlessly, to achieve their aims. The bank clerk, the journalist or the assembly line worker did not propose to senior managers how their daily work could be made more efficient by the adoption of new technology! Later in this chapter we will look at how surveying firms should go about identifying, acquiring and employing new technology in the workplace.

Is construction unique?

Is it reasonable to compare the property and construction industries with the banking, newspapers and automotive industries?

Banking, just like construction and property, is a service industry; however its product is less tangible than a building and is also much more standardised, in that the same products will be offered as a service nationally throughout the bank's marketplace. Banks are also very profitable, and are large commercial organisations employing thousands of staff. In addition, they have a large base of customers and principally deal in financial data. It was the computer power of storing, sorting and manipulating data that led to the original introduction of computers into the banking industry. Therefore, because the industry had large profits to reinvest in their business activities, a standardised product, hundreds of thousands of data sets and a very structured data format, it was very easy for new technology to be introduced.

The newspaper industry introduced new technology for a different reason – introducing new working practices. Here the problem was archaic working practices that had changed little from the days of William Caxton, together with very powerful trade unions that were opposed to the introduction of the new technology. The trade unions were acting in what they thought was their members' best interests in trying to protect the livelihoods of their members. However, in refusing to acknowledge or even consider any new working practices, they eventually succeeded in making the skills of their members irrelevant to the modern newspaper industry. New technology was introduced by being championed by the newspaper owners, and this led not only to new working practices but also to new products. The lesson here is that the advance of technology cannot be ignored or reasonably repulsed, and it is inevitable that some traditional skills will become redundant in the process. Furthermore, to be successful, the introduction of new technology requires a champion at senior management level. Once again the newspaper industry is an example of an industry comprising very large organisations with high levels of profitability and a large workforce.

The introduction of new technology into the automobile industry was for different reasons again. First, the aim was to integrate design and manufacture via CAD/CAM and robotics; secondly, it was to allow new working practices associated with supply chain management. Here, once again, the introduction of new technology led to the development of new ways of working. Senior management again championed the introduction of new technology, and the firms in question were profitable, had huge annual turnovers, employed thousands of staff, and manufactured very similar products in vast quantities under factory conditions.

These examples from other industries demonstrate that, to introduce new technology into the workplace successfully, it is necessary to:

- Champion new technology at a very high level in the company.
- Ensure that the new technology is focused upon helping the firm achieve its business objectives.
- Re-engineer existing business processes, often leading to the introduction of new ways of working, new services, and the redundancy of some traditional skills.

The construction and property industries, on the other hand, have quite different characteristics (see Table 6.1) in that they are typified by:

- A proliferation of small firms forming temporary alliances upon a project basis for the duration of that project.
- The separation of the design and construction activity.
- Uncertainty due to variable demand, on-site construction hampered by inclement weather and the lack of a standard product.
- Fierce competition for work that creates low levels of commercial profitability.
- Each product being, to a greater or lesser extent, unique.
- Hundreds of stakeholders in any one project, many of whom may have conflicting business objectives.
- Displaying considerable resistance to any change in working practices.

Table 6.1 Characteristics of the construction and property industries

Characteristic	*Example*
Fragmentation	<ul><li>Design and construction processes separated</li><li>Any one project has many stakeholders</li><li>Industry structured upon basis of project teams that are temporary alliances</li></ul>
Uncertainty	<ul><li>Inclement weather can adversely affect progress on site</li><li>Very competitive environment</li><li>Process accommodates design changes during the construction phase</li><li>Every building is unique to a lesser or greater extent</li><li>Buildings are becoming increasingly more complex</li><li>Nomadic labour force</li></ul>
Poor communications	<ul><li>Missing or conflicting design information at the construction stage</li><li>Late transfer of information between design team and the constructor and constructor and subconstructors</li><li>Conflicting information sources, for example between bills of quantities and project specification.</li></ul>

Comparisons between the construction and property industries and those of banking, newspaper publishing and the automotive industry are therefore not entirely reasonable, but should not be completely dismissed. The principal differences are that the property and construction industries lack a champion to lead the introduction of new technology, and that the average construction and property firm is very small – typically employing fewer than 12 staff – as well as unprofitable, many achieving less than 3 per cent profit margins. A further complication is the organisation of the construction industry on a predominantly project basis, with a proliferation of stakeholders, which leads to poor communications, misunderstandings and errors. Where then might a champion be found? Given the poor success rate of acceptance of technology into the construction and property industries, you would be forgiven for thinking that none exists.

The most obvious champion is the senior management of any large organisation active in the construction and property industries, such as a large contracting organisation or a multi-professional practice, and there are indeed examples of this. The new technology in these instances is usually associated with company based information systems introduced to achieve company objectives. These successes are due to the fact that they are contained within the boundaries of one company or organisation, where common standards can be introduced and enforced, and that those large companies are more likely to have both the vision and the funding to support the appropriate introduction of new technology. Typically, however, these instances do not display the characteristic of introducing new working methods, but are targeted upon the achievement of a company's business objectives. The other obvious champions are the larger clients – those who are constantly commissioning building works. Most prominent of these is central government. The government has commissioned various reports, e.g. Rethinking Construction, and funded initiative such as Constructing Excellence in the Built Environment into the operation of the construction and property industries, most of which have recommended greater use of new technology allied to the introduction of new working practices. Interestingly, unlike the previous example these have been targeted at the project level and are geared to achieving the client's objectives. Various government initiatives to introduce the widespread use of web-based systems for procurement, discussed in Chapter 5, have an example of this approach.

New technology and information

Just because something is technically possible it does not mean that it should be adopted in the workplace, and there are many examples of new technology being employed for its own sake. Such instances are doomed to failure, as they are not allied to supporting the achievement of the company's business objectives. New technology is often referred to as information technology; note that the term 'information' precedes that of 'technology'. This is indeed the correct relationship, as new technology or information technology should always support the company's information systems. It is the information systems that are important, more so than the supporting technology. The purpose of any information system is to ensure that appropriate accurate information is made available to the correct person timeously in order that a decision can be made. Information itself is merely processed data, and data are facts that in themselves have no apparent meaning. The problems associated with the building of information systems revolve around who needs what information when, how to acquire appropriate data, and how those data can be processed into the requisite information.

Information systems are the product of system analysis, and it is the systems analysis that leads to the business re-engineering and new ways of working described earlier in this chapter. Systems analysis concerns itself with the optimum way of achieving a task, and not the recording of how a current task may be being performed. It is the missing information systems that have led to the many failures in investment in new technology in the construction and property industries. This is also partly responsible for the failure to adopt new working practices, as most investments in new technology-based information systems have been based upon the automation of existing working practices. This has resulted from the bottom-up approach taken by many investments in new technology in the construction and property industries, where investments have been made to support individual information needs rather than those of the company overall. It is also further compounded by the identification that, in reality, two mutually supportive information systems are required to assist build environment businesses, namely:

1. A company-based information system to support the company's objectives
2. A project-based information system to support the client's objectives.

This in turn leads into the very important concept of the need for integration of information systems in the construction and property industries. Earlier in this chapter fragmentation of the industry, which is typified by the large number of stakeholders involved in any one project as well as the preponderance of islands of computing or islands of information, was identified as problems to be overcome. The obvious solution to these problems is integration. Unfortunately, the adoption of new technology does not automatically lead to the integration of new systems.

Degrees of integration

Earlier in this chapter it was identified that some of the barriers to effective and efficient built environment information systems are the fragmented nature of the built environment and information system islands. The obvious solution is to adopt integration, but what should or could be integrated?

The traditional separation of design and construction activities is one obvious area where integration should be sought, particularly since this is seen as being the principal cause of poor industry communications. There are, however, new procurement systems that integrate these activities, for example, design and construction, but these, even when they occur within one commercial company still have information system problems associated with communication between the designer and the builder. Often the 'design and build' organisation is a federation of separate companies or even a virtual organisation that is in effect little different from the traditional procurement routes where design and construction are disparate activities. Indeed there is little evidence of new working practices being employed to support the design and build procurement processes. Integration therefore is not as simple as adopting design and build as a means of procurement.

Integration could be achieved at numerous different levels within the built environment, e.g. individual, departmental, organisational, etc., each of which would result in the attainment of differing benefits. The degree of integration achieved can also differ and could vary from low to high.

Considering the individual level first, integration at this level is likely to be based upon the integration of personal data and information systems with those of others. The advantages of adopting integration here would be to help eliminate islands of information and to empower teamworking and it would probably also be beneficial in terms of personal efficiency gains. The reasons why an individual may seek integration at the personal level are many and varied, but could include to stay in employment, to increase their earning potential, to extend their personal standing and authority in an organisation, to gain new skills and change employment and possibly even to create a new job (become the office guru and thereby become indispensable).

There are numerous examples of this type of integration having taken place in the built environment especially in terms of the information systems used by professional people, such as architects, surveyors and engineers, e.g. commonly today CAD systems are project team collaboration tools rather than individual drafting tools as are bills of quantities production systems.

The next possible level of integration is at the departmental level. Here the aim should be to integrate information system models, for example, the integration of multiple information systems used by one department. Once again the benefits are likely to be efficiency related and the reasons for wanting to integrate at this level are similar to those identified at the personal level and include departmental survival, increase profitability of the department and thereby departmental importance/ranking, assume additional departmental roles/duties and responsibilities and lastly to possibly create new roles with the department (organisational empire building). Examples of this degree of integration are more difficult to find in the built environment due to the professional nature of many firms and their need to be organised upon a project basis.

Integration at the organisational or company level is the next level of possible integration. Here the objective should be to integrate corporate knowledge and/or to focus the information systems upon the companies business objectives. Integration here is also likely to be related to the integration of multiple departmental and/or company-based information systems. The aim here being to achieve some form of competitive advantage in the marketplace and thereby increase market share. The reasons why integration should be undertaken at this level are similar to those already identified previously but are obviously enterprise related and include company survival in the marketplace, to increase profitability and or market share, to enter new markets and/or create new markets for your services.

So far all of the levels of integration have been enterprise- or company-related, but the next possible level of integration is at the project level. Integration here comprises of integrating all of the project stakeholders information systems to have a project focus enabling overall project success in terms of the clients objectives and possibly beyond this to encompass the whole project life cycle. This is an area that is currently receiving a great deal of attention in the built environment.

The pinnacle of integration is found at the industry or national level. Here the objective is to integrate all project and industry knowledge and information. The benefits being sought here are to improve the performance of an important national industry and also if possible to expand that industry expertise into international markets. This is another topical subject that has been the focus of many government-funded reports and initiatives and remains so today. Three such initiatives were 'Construct IT – Bridging the Gap' (Anderson Consulting, 1995), 'Building IT 2005' (Construct IT Forum, 1996) and 'Technology Foresight: Progress through Partnership 2: Construction' (Cabinet Office, 1995).

Integration and new technology

New technology is something of a paradox when considering how it can be employed to achieve integration of information systems in the construction and property industries. The simplest route is where everybody uses the same computer operating systems and software. However, even this route can lead to problems if people are using different versions of the same software, have set up the software to operate with different default settings or, in the case of CAD, simply adopted contradictory layer arrangements. This can of course be simply remedied by enforcing a framework or structure regarding how the new technology should be employed. This requires a champion to drive the concept forward and ensure that the framework is adhered to. Examples of this approach to integration based upon new technology are most often found within one firm or a single organisation, and consequently this approach is well suited to the company-based information systems identified earlier in this chapter.

A more common instance is where integration is required across a range of operating systems and a variety of disparate software. The solution here lies in the adoption of either neutral file formats or open architecture computing. Neutral file formats have long been the Holy Grail sought by organisations and governments as the tool that will enable easy transfer of information between differing computer systems, especially CAD systems. Early attempts at solving this problem were based upon translators that were used to convert data from one system to another and vice versa. This was not really a feasible solution as the number of translator applications required grew exponentially to the number of CAD systems in use, and it was obvious that another solution was required. In Europe this took the form of the Initial Graphics Exchange Specification (IGES) and, later, the Standards for the Exchange of Product Data (STEP). STEP, now an international initiative, is concerned with the exchange of product data between computer-based information systems. In the USA, a similar initiative took the form of the Product Data Exchange Specification (PDES), which in turn led to the development of Product Data Modelling (PDM). PDM is yet another international initiative concerned with the use of new technology for the representation and exchange of product data. It is based upon four other international standards: Electronic Data Interchange (EDI; discussed in Chapter 6); ISO 8879 – Standard Generalised Mark-up Language; ISO 10303 – STEP; and ISO 13584 – Parts Libraries.

The most recent initiative is that of Computer Acquisition and Lifetime Support (CALS). This has its origins in the US defence industry and is now also a UK government-sponsored initiative, and includes the construction and property industries within its scope. CALS is also a combination of existing and emerging standards, including EDI and STEP. All of these initiatives are government sponsored, are targeted at achieving integration at the data level, and are not specifically targeted at the construction and property industries but at all industries – especially manufacturing industry.

Given the predominance of small to medium-sized enterprises (SMEs) in the construction and property industries, and the complexity of the solution in relation to the type of project that constitutes the day-to-day workload of these firms, it is obvious that the generic solution to data integration does not lie down this avenue. A more likely solution, and one that is commonly adopted by SMEs, is that of adopting the proprietary neutral file formats that are available commercially in the new technology domain. These provide only a partial solution to the problem of integration, but are generally free, commonly used, and reasonably well understood by users; however, they cannot really be compared to the former solutions in scope, intent or usefulness. Examples include the use of a text file and HTML files for text-based documents, comma-delimited files or space-delimited files for numeric data, and DXF, the AutoCAD neutral file format, for CAD files. Although readily available, and generally free, they provide an incomplete and imperfect solution.

An alternative solution that is showing promise is that of industry foundation classes (IFCs), which are being developed by the International Alliance for Interoperability (IAI). This is an international non-profit making association of companies and researchers active in architecture and engineering construction. The IFC development workload has been devolved to the members on a national basis, each nation being set an objective that contributes to the overall international goal. IFCs work at the project data level and are associated with object oriented programming, which is an entirely new and radical approach to the development of computer programmes and CAD systems in particular. In simplistic terms, the basic object oriented concept is that the world is made up of objects, that these objects can inter-react with each other, and that they can be simulated in the software programming. Consequently, object oriented CAD software would be able to create dynamic models, or models that can react to their environment or usage. Integration is at the core of the IAI initiative, and IFCs are merely an attempt to define objects in the construction and property industries, which can then be employed by the software designers to develop software for the industry that is capable of displaying 'interoperability' or integration. Software built to comply with IAI IFCs probably stands a much greater chance of success and adoption across the industry than STEP or CALS.

The last possible solution to the problem of integration is based upon 'open systems computing'. This relates to the desire for hardware platform and operating system independence, and is commonly interpreted today as meaning internet technologies. Many of these tools are based upon the standard generalised mark-up language (SGML), which is in effect a set of rules for constructing other internet languages. There are many of these and they are constantly under development, so many cannot yet be considered mature. The most common example is hypertext mark-up language (HTML). HTML, although useful as a publishing medium, has limitations for business use, and amongst its faults are the fact that the designer has little control over what the finished page looks like on the computer screen, and also that it is really only useful for text-based information. A more valuable example for the built environment is virtual reality mark-up language (VRML), which is an internet standard for 3D graphical information. This enables the viewing of 3D graphical models over the internet.

Extensible mark-up language (XML) is a simplified form of SGML, and is another set of rules for creating mark-up languages. XML-based languages, however, are not concerned with appearance, but rather contain information about the logical structure of the document. The important thing about XML-based languages is that they more readily enable the transfer of documents between computer systems without those documents losing their structure. XML could be used, for example, to send an invoice between one computer and another, despite both computers having different hardware architectures and operating systems. Indeed, XML shows every promise of more readily and easily enabling electronic data interchange (EDI) to be achieved than do current technologies.

The solution to the built environment problem of integration may well lie within a combination of these internet technologies.

New technology and the knowledge worker

Few people in the construction and property industry actually earn their living by physically creating a solid product, such as the building itself or any of its many components. Most jobs instead involved the management or processing of data, and people working in these fields are commonly called information workers. Information workers can be further subclassified as data workers, as people who store, retrieve and manage data, or as knowledge workers – those who create new information or knowledge. Professionals within the built environment are knowledge workers because they add value to the product via the application of their knowledge and expertise in order to create new knowledge about the product – for example, the architect produces the conceptual design, and the quantity surveyor the cost and procurement advice. New technology is capable of supporting both the data worker and the knowledge worker.

The object of providing new technology for knowledge workers is that by doing so their productivity will be improved. This is of prime importance to most professional firms, as the knowledge workers are also the fee earners. Any investment in new technology here should aim to integrate the knowledge and expertise of the knowledge worker into the business, rather than create a desktop tool for the sole use of the individual fee earner. Once again integration with other office systems is all important, and this is normally interpreted as being the easy transfer of information between knowledge work systems or data stores. This invariably necessitates the use of networks and/or intranets, and perhaps also the internet. Examples of knowledge work systems in the built environment include computer-aided design packages (CAD) for architects, and estimating and bills of quantities production systems for quantity surveyors. Other examples include modelling and rendering software and analysis software (e.g. structural or environmental analysis software). It is worth noting that these tools in themselves are incapable of doing anything very much, and it is only through the application of the workers' skills, expertise and analysis of the results produced that new knowledge or information is created.

Recent trends in built environment knowledge work systems display this desire for the systems to be integrated into the business and/or the project environment rather than being seen as a desktop tool for the sole use of one person. ArchiCAD V9.0 is marketed as being a network-based project design tool that can be used by a team of designers to develop and manage a project's design and documentation. This tool goes way beyond the normal concept of CAD as being an electronic drafting board. For example, Q-Script, a tool for the quantity surveyor, displays similar characteristics along the lines of enabling collaborative project working upon a team basis and, being a networkable product, also enables the management and documentation of the surveying services on a project.

Other new technology tools, although not directly helping the productivity of the knowledge worker, can be employed to improve collaboration and communication between workers. These tools are concerned with enabling communication and collaboration generally, and are not environment specific. Examples of these include electronic group diary and schedule systems, contact management systems, document management systems, company intranets and e-mail systems, publishing systems, and even simple bespoke databases and word processed documents. All of these can be used to share and distribute knowledge. However, as before, investment in these must be appropriately focused upon the company's business objectives in order for any real benefit to be realised. The increased productivity of any of these systems does not in itself guarantee increased productivity and efficiency. E-mail systems and the internet in particular can be considerable time wasters if not properly integrated and controlled in the workplace.

New technology and graphical information

Graphics play a very important role in both communication and the transfer of information in the property and construction industries. These graphics can take many forms, such as freehand sketches, detailed technical drawings, and as built drawings. Traditionally this graphical information has been passed between project stakeholders in the form of 2D drawings. Techniques have even been developed to represent 3D views of buildings on a 2D page (e.g. isometric and oblique representations). Even relatively small projects can result in the production of tens or even hundreds of drawings.

The project drawings are merely a proven method of recording and communicating project information. The conventions for the production of project drawings have developed over many years but, despite having been in use for many generations, their production is not problem free, even today. The production of project drawings is a collaborative task, with many different individuals and firms contributing to the final product. Almost without exception, these drawings are produced today using a computer-aided design (CAD) system.

The backbone of all CAD software is a database, although most users are not aware of this fact. The structure of a CAD database is different from that of a normal relational database such as MS Access, which is made up of fields, records and tables. A CAD database is just a well organised list of data, and is capable of storing two kinds of information: the geometry information required to produce the CAD drawing or model and non-graphical information inserted by the user. The technique for inserting non-graphical information into a CAD database is to add the information to blocks via the block attributes facility (Figure 6.1). Historically, designers have not commonly adopted this facility, since it is perceived as being non fee earning work and for the benefit and use of other professions, and has therefore largely been ignored. The advent of object oriented CAD software will to some extent solve this problem, in that much of the data in an object oriented CAD database will be produced automatically without the need for the CAD user to insert the information manually.

However, if CAD is to form the basis of a project information data model capable of supporting collaborative working, the capability of the CAD database to store and retrieve non graphical information must be used to its full potential. How, then, might

CAD block attributes be used to store project data in the CAD database? The most obvious application is the recording of component attributes (e.g. information associated with doors, windows, trusses, etc.). Block attributes can also be used to store pre-design information associated with the project brief, such as information concerning an employee workstation or office in an office development. Indeed the opportunities are only limited by imagination and ability, and can include the incorporation of CAD data into reports, specifications, schedules and other construction-related documents (Figure 6.2).

Block attributes can usually be displayed on the CAD drawing that is being worked on, or extracted into a data file that can be subsequently imported into another software application, such as a spreadsheet or a word processor. More importantly, block attributes can also be linked to a field in external relational databases via the adoption of Structured Query Language (SQL). An example of this technique would be to link a door block in a CAD drawing to an external database that contained other fields relating to door dimensions, types, hardware, etc., thereby enabling the production of the project door schedule. This leads to the concept of the CAD drawing becoming an intelligent interactive drawing that does not require the CAD user manually to insert all of the project data associated with block attributes. It also exemplifies the basic computing rule of only ever having to insert data into a computer system once.

CAD is therefore capable of supporting project information databases. All projects have a need for information management, and any supporting information system must be focused on the following

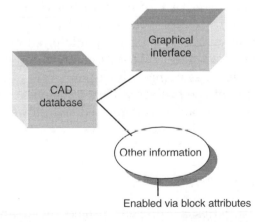

Enabled via block attributes

Figure 6.1 CAD structure.

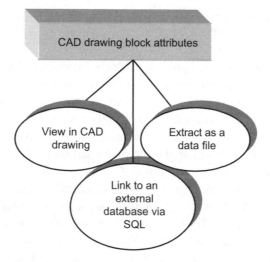

Figure 6.2 CAD block options.

project-based critical success factors (CSFs), all of which are interdependent:

- Timely completion of the project
- Completion of the project within budget
- The quality of the finished product fully meeting the functional requirements of the client.

A core objective of any information system is to ensure that the correct information reaches the appropriate person when it is needed. In the construction industry, communication lines are often regulated by the form of contract adopted rather than by the optimum information system architecture. Any built environment, project based information system must be capable of supporting all three CSFs of time, cost and quality. Traditional built environment, project based information systems only support one of the core CSFs; that of completion of the project within budget.

Project information can be categorised as follows:

- Commercial
- Technical
- Managerial.

Each category satisfies a different project need and has unique characteristics. A project based information system would need to accommodate all three formats of information and recognise that administrative functions are interwoven with commercial, technical and managerial information needs.

Commercial project information (e.g. accounts, payroll, tax returns, invoices, labour returns, plant returns, delivery notes, etc.):

- Needs to be 100% accurate
- Is not time-critical (in terms of project CSFs)
- Must be 100% auditable
- Is historical in nature
- Can be used as data to predict future events (project control)
- Is largely administrative.

A CAD-based project information model has little to offer in terms of support for commercial based project information, which in many respects reflects more the requirements of a company based information system. The information technology tools best able to support this model lie outside the functionality provided by a CAD system even when linked to other systems such as a relational database.

Technical project information (e.g. client briefings, design guides, project drawings, project specification, bills of quantities, engineering calculations, environmental impact analysis, contract conditions, variation orders, etc.):

- Needs near 100% accuracy
- Is approaching 100% time critical
- Should be auditable
- Is procedural driven
- Is directly related to project control mechanisms (CSFs)
- Results in a project 'knowledge base' (expertise and experience).

This category is the domain in which a CAD-based project information system has the most to offer, even without the advent of OOPs-based CAD software. The techniques involved were discussed earlier in this chapter, and their application is limited only by imagination and technical competence. Research in this area is being targeted at the development of intelligent CAD software that guides the user through technical or professional processes and integrates either technical or professional knowledge with the CAD software. An example of this would be a CAD system that advises on fire escape regulations as the design develops, or the building regulations in respect of a stair design.

Managerial project information (e.g. project time, cost, quality, feasibility factors):

- Requires a degree of accuracy
- Is concerned with trend analysis
- Has no audit requirement
- Has a significant project control requirement
- Is 100% time-critical
- Is forward looking
- Is management based.

Much research is currently being devoted to the development of CAD-based information systems that are capable of supporting managerial project information, an example being the linking of CAD models to project based planning systems. Most of these initiatives fall under the umbrella of Computer Integrated Construction (CIC). There are also a number of examples of this use of new technology being introduced into the workplace; however, these are normally in instances where design and construction fall under the umbrella of one company, e.g. a design and build contractor, or a multi-professional firm.

CAD-based project information systems are therefore most able to support the project requirements for technical information systems, and show some promise in being able to interface with managerial information systems, especially in circumstances where there is less project fragmentation. Yet again they are based upon the integration of information and systems, and to achieve fully the benefits on offer will require the development of new working practices, as was the case in the newspaper industry.

New technology and commercial advantage

Can the adoption of new technology assure commercial advantage?

The adoption of new technology is often though to be associated with commercial advantage, and there is a huge volume of published material on the subject. One thing is sure; what is done today (should it prove successful), the competition will copy tomorrow. Consequently, commercial advantage associated with new technology is often short lived. Companies in the built environment are also difficult to differentiate from each other; they all provide a similar range of services, the setting up of new firms is not prohibitively expensive or difficult, and none have the authority actively to influence their own supply chains. These characteristics, along with the high level of fragmentation in the industry, make it very difficult to adopt new technology with the aim of achieving any lasting commercial advantage. Furthermore, a large number of SMEs active in the built environment do not have the time, resources or vision to investigate or experiment with how technology could be adopted to gain any commercial advantage. A more common scenario is that technology has to be adopted to enable them to compete with rival firms that have already attained some advantage. Motivators of this type are unlikely to result in either the successful introduction of new technology into the workplace or the attainment of the expected benefits.

Commonly, new technology is adopted to achieve efficiency gains and thereby a cost advantage over competitors. Other businesses, fewer in number, have adopted technology to develop new services that enable them to differentiate their services from rivals. A few have adopted technology to enable the adoption of new working practices or structures, and it is perhaps these that will prove to be the most enduring and valuable. The successes achieved by the banking, automotive and newspaper industries all resulted from the adoption of new technology allied with new working practices and organisational structures. The adoption of new technology itself is therefore unlikely to result in any lasting commercial advantage, although failure to adopt technology commonly used by competitors could lead to a failed business. The most likely way in which technology could be adopted to achieve lasting commercial advantage would be where it is adopted to support new services, to enable the restructuring of the organisation, and/or to introduce new working practices.

New technology and people

For new technology to be welcomed into the workplace, the users must accept it. To be accepted by the users, technology must be seen as benefiting them by making their jobs less tedious or giving users new skills and responsibilities. It is also wise for the potential users of a system to be involved in its selection, development and introduction, as this engenders a feeling of ownership. Any attempt to impose a system upon users unilaterally is unlikely to prove to be a rewarding experience.

The introduction of any new system into the workplace will require staff training. Training is vitally important, and many new
technology systems fail to produce the expected benefits owing to a
lack of formal training. It should not be seen as an afterthought but
as an essential ingredient, and it is quite likely that training will be
required over an extended period of time, rather than just a few
days. It is also good practice to have staff train each other, and for
the user knowledge to cascade down through an organisation.
There should always be at least two people with intimate knowledge of any one system to ensure continuity of expertise in the
event of illness or staff changes. Initially training must take place
away from the day-to-day workplace pressures, although latterly it
can be integrated into the daily routine.

The introduction of any new technology system will inevitably
result in a fall in output as staff become familiar with the new systems, and it is likely to be some months before performance recovers to its former levels, and perhaps a year or more before any
productivity benefits become apparent. Staff should work their
way through the performance levels, from becoming familiar with
the technology to becoming competent users and ultimately to becoming innovative with the technology. Many firms fail to
progress beyond the stage of becoming competent with the technology. It is a good idea for each member of staff to become the
guru with regard to any one application and then to act as a focal
point for queries relating to that application.

Training is not free, and it is not unusual for the cost of training, including the associated fall in staff productivity, to exceed
the investment made in purchasing the new technology itself.
Training should also form part of each system upgrade, and not
be provided only when the system is first introduced into the
workplace.

How best to employ new technology

The stimuli for employing IT in the workplace are not always positive in nature, and are often associated with what clients expect of
your firm, the need to collaborate with others in a specific project
team, staff expectations, or simply the need to remain competitive.
As identified earlier, investments made upon these bases are unlikely to lead to success. It is not the IT itself that is important, but
how it is employed within your organisation that is vital.

In giving advice upon how IT should best be employed in an organisation, consideration has to be given to these factors:

- The size of the firm (SME or other)
- Whether the IT is to service business focused information systems or project based information systems (or both)
- Whether the IT investment is to provide knowledge work system tools.

A distinction must be made between the resources and capabilities of SME and larger organisations. SMEs do not usually have individuals within the organisation with the necessary IT skills, knowledge or vision to identify IT/IS opportunities within their firm, and neither do they have the capacity to service and maintain their own IT infrastructures once installed. Furthermore, it is these firms that are least likely to be able to absorb the consequences of any IT investment failure, and they do not usually have the resources or the inclination to employ the requisite professional IT/IS advice upon a consultancy basis. Given that the vast majority of firms in the built environment fall into the SME category, the advice in this section of the chapter is directed towards their needs. There are two basic categories of information systems in the built environment.

1. Those that are focused upon the firm's business objectives (e.g. to acquire more clients or increase fee income)
2. Those that are targeted at the project or client's objectives (e.g. to complete the building on time, to cost, and to the expected level of quality).

Project based information systems are geared to enabling collaborative working practices and establishing project databases. IT is required to support both types of information system. A third role for IT in the built environment is to provide knowledge work systems for knowledge workers, as identified earlier in this chapter. This advice focuses upon the use of IT to support business objectives via company-based information systems or knowledge work systems.

Experience proves that *successful* IT investments have the following characteristics:

- IT investment is linked to core business objectives
- Successful IT strategies are driven from the top down in organisations, and should be championed by a senior partner/manager

- IT investments should be linked to an IT strategy: that strategy should cover a period of 4–5 years and itself be subject to periodic review
- Users need to be involved in the selection and introduction of any new IT system to ensure that they have a sense of ownership of the system
- Successful introduction and acceptance of any new system must be supported by adequate and frequent staff training
- IT systems require effective user support and maintenance.

Other experience has shown that *unsuccessful* IT investments are characterised by:

- The lacking of integration with other IS/IT systems, thereby creating islands of computing
- Being technology driven (technology for technology's sake)
- Inadequate provision of IT equipment (e.g. the sharing of computers and software by users).

IT investment best practice

The methodology shown in Figure 6.3 is recommended to SMEs with regard to IT investment.

First, three tasks need to be undertaken concurrently: two if not all three of these can be performed internally with some guidance. The tasks are:

1. To survey existing IT facilities
2. To survey existing information systems
3. To identify the firm's business objectives.

The survey of the firm's existing IT infrastructure identifies the investments that have already been made, the outline specification of the equipment, the users, and the tasks performed. The IT survey should also include the definition of the firm's existing IT policy, which should cover very basic factors such as file-naming conventions, standard directory structures, and staff acceptable use policy. An acceptable use policy sets down the rules and regulations regarding how employees may or may not use the information technology tools provided in the workplace. The intention is to avoid, and hopefully prevent, any malfunction of the company's supporting

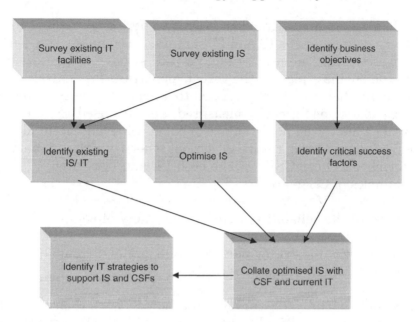

Figure 6.3 IT development methodology.

information technology due to the behaviour of employees. The survey should also identify the firm's back-up policy and procedure.

The objective of the IS survey is to identify how information flows through the firm and where and how it interfaces with the IT currently in use. This is the most difficult of the three initial tasks for the layman to undertake. It would normally be performed by a systems analyst; however, with a little background reading and some simplification, it is possible for this task to be performed in-house. The approach is to identify the tasks performed and where the data/information required to complete each task is created, stored, updated and deleted. The survey should also identify where and how the existing IT interfaces with the IS. The IS survey data should be analysed to identify core, duplicate and redundant systems, and to identify the IS/IT development opportunities. Ideally 'clusters' of activity should be identified, and it is these clusters that will probably determine the best opportunities for the development of new IS – and equally, the existing ISs that are little used or of dubious value.

The firm's business objectives would normally be contained within the firm's business plan. A business threats/opportunities analysis (SWOT analysis) can be used as a vehicle to clarify and focus the business objectives with the firm's partners if the business plan is weak or non-existent. All of the business objectives should then be investigated to identify where they related to the IS/IT as defined in the IS and IT surveys. To complete this part of the exercise, critical success factors (CSFs) have to be developed from the business plan or the threats/opportunities analysis. CSFs are factors that are directly related to a firm's business objectives, are capable of being easily measured, and are used to determine whether the desired performance is being achieved and to identify where failure would result in serious shortcomings in business performance. Often CSFs will be both IS- and IT-related.

Examples of CSFs are:

- Improve the reliability, security and safety of computerised information
- Provide the means for partners to assess and compile fee bids from historical company records.

The results of the IT/IS surveys and the CSFs are then correlated to identify and prioritise IT opportunities. It is the IS requirements of the business objectives and/or the firm's IS system that should drive the identification of the IT facilities required. A number of alternative IT strategies should then be developed and subjected to a discussion by the firm's partners. An IT strategy is simply a prioritised shopping list that is costed and linked to a timescale, normally a period of 4–5 years. Once an agreed strategy is developed, all investments should be made in accordance with that strategy. The strategy is usually subject to annual review but not wholesale change within the period of its life.

There are a number of related management issues that should be borne in mind at this stage:

- The acquisition of new technology should not lead to an associated increase in the firm's non-productive overhead costs. These can be minimised by adopting proven simple technology that is easily managed and maintained. Consideration should be given to acquiring a maintenance contract with a local IT consultancy for maintenance and repair of the IT systems.
- When purchasing equipment, seek turnkey contracts and quotations on a performance specification basis. Most SMEs can define what they want the system to do but are unable to specify or assess adequately technical tenders to support that requirement. Turnkey contracts also place the onus on the contractor to leave a fully working system in place upon completion of the contract.

- Actively seek out hardware suppliers who provide extended warranties with their equipment. Three year on-site warranties are now commonplace.
- Ensure that any investments made in software are with firms that have a good track record of continuous research and development, and are likely to be in business for some years to come.
- Be aware that even off-the-shelf software requires customisation for use in the workplace.

Conclusion

New technology should be seen by quantity surveyors as an opportunity rather than a threat. However, any investment needs to be based upon sound business practices and should always be related to the firm's business objectives. Furthermore, investments should be related to an IT strategy and championed by a senior member of the organisation. It is unlikely that new technology will provide a firm with any lasting commercial advantage, but failure to invest could lead to the firm being unable to compete with its rivals. Surveyors will be required to adopt and use new technology tools that enable collaborative working on a project basis, and it is quite likely that these systems will be based upon internet technologies. To gain the maximum benefit from any of these new technologies, surveyors are going to have to adopt new working practices and probably also develop new skills.

Bibliography

Betts, M. *et al.* (1999). *Strategic Management of IT in Construction.* Blackwell Science Ltd.

Castle, G. (2000). *Planned IT Infrastructure: Napier Blakely Winter.* Building Centre Trust.

Websites

The IT Construction Forum, http://www.itconstructionforum.org.uk/

The Construction Industry Computing Association, http://www.cica.org.uk/

Constructing Excellence in the Built Environment, http://www.constructingexcellence.org.uk/

The Building Centre Trust, http://www.buildingcentretrust.org/

7

Global markets – making ends meet

Introduction

This chapter will examine the role of the quantity surveyor in the global marketplace, together with the problems faced by an organisation wishing to expand into European and/or global markets. Also included is an explanation of existing and proposed EU public procurement legislation and the procedures that must be followed to comply with European public procurement policy. During the last 5 years, the RICS has increased its global presence including the appearance of several European and US universities and higher education establishments in the RICS partnership programme. However, all eyes are on China, with an economy growing at 9 per cent per annum (compared with the average for the Euro zone of 1.8 per cent) and a 26 per cent year-on-year increase in infrastructure investment, such as bridges, factories and power plants.

The multi-cultural team

The *Le Monde* cartoon featured in Figure 7.1 illustrates most quantity surveyors' perception or even experience of working in or with multi-cultural teams. Quantity surveyors have proved themselves to be adept in a diverse range of skills, often over and above their technical knowledge, with which they serve the needs of their clients. However, when operating in an international environment these skills and requirements are complicated by the added dimension of a whole series of other factors, including perhaps the most influential – cultural diversity. Companies operating at an international level in many sectors have come to realise the

Figure 7.1 Making ends meet (Source: Serguei, *Le Monde*).

importance of a good understanding of cultural issues and the impact that they have on their business operations. In an increasingly global business environment in which the RICS is constantly promoting the surveyor as a global player (for example, the RICS Global Manifesto), it is a fact that the realisation of the importance and influence of cultural diversity is still lacking in many organisations seeking to expand their business outside the UK. Figures produced by the FIEC show that in 2004, amongst the EU states, France, Germany and Sweden exported the largest percentages of national construction turnover to world markets (Figure 7.2). What is it that makes these three countries particularly successful? Is it their approach to the construction process or their multi-cultural attitude that enables them to transcend national boundaries?

Today, international work is no longer separated from the mainstream surveying activity; EU Procurement Directives, GATT/GPA (Government Procurement Agreement of the World Trade Organisation), etc., are bringing an international dimension to the work of the property professionals. Consultants from the UK are increasingly looking to newer overseas markets, such as Europe and regions, where they have few traditional historical connections, such as South East Asia and China. Consultants must compete with local firms in all aspects of their services, including business

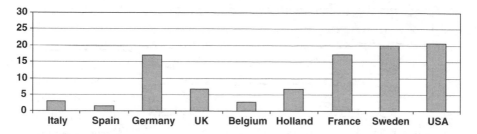

Figure 7.2 Construction exports in terms of billion € 2004 (Source: European International Contractors).

etiquette, market knowledge, fees but above all, delivering added value, in order to succeed. With the creation of the single European market in 1993, many UK firms were surprised that European clients were not at all interested in the novelty value of using UK professionals, but continued to award work on the basis of best value for money. As pointed out in Chapter 1, the UK construction industry and its associated professions have ploughed a lonely furrow for the last 150 years or so as far as the status and nature of the professions, procurement and approach to design are concerned, and it could be argued that this baggage makes it even more difficult to align with and/or adapt to overseas markets. Certainly companies like the French giant Bouygues, with its multidisciplinary *bureau d'étude,* have a major advantage in the international markets because of their long-established capability to re-engineer initial designs in-house and present alternative technical offers. Unlike the UK, contractors in France are the largest employers of construction professionals, which gives them the capability to analyse design/construction problems in order to arrive at the best value added solution, before submitting a bid to the client. Outside Europe, the USA, for example, has seen a decline in the performance of the US construction industry in international markets (see Figure 7.2), a trend that has been attributed in part to its parochial nature in an increasingly global market. Internally, strong trades unions exercise a vice-like grip on the American construction industry, to the extent that in some states access to construction sites, even for casual visitors, is forbidden. The same trades unions regularly identify organisations that use non-union labour by parking a giant, portable, inflatable 'black rat' outside their headquarters – behaviour not seen in the UK since the 1980s.

The construction industry, in common with many other major business sectors, has been dramatically affected by market globalisation. Previous chapters have described the impact of the digital economy on working practices; multi-national clients such as Coca Cola and BP demand global solutions to their building needs, and professional practices as well as contractors are forging international alliances (either temporary or permanent) in order to meet demand. It is a fast moving and highly competitive market, where big is beautiful and response time is all important. The demands placed on professional consultants with a global presence are high, particularly in handling unfamiliar local culture, planning regimes and procurement practice, but the reward is greater consistency of workload for consultants and contractors alike. In an increasingly competitive environment, the companies that are operating at an international level in many sectors have come to realise that a good awareness and understanding of cultural issues is essential to their international business performance. Closeness and interrelationships within the international business community are hard to penetrate without acceptance as an insider, which can only be achieved with cultural and social understanding. In order to maintain market share, quantity surveyors need to tailor their marketing strategies to take account of the different national cultures. Although some differences turn out to be ephemeral, when exploring international markets there is often a tightrope that has to be walked between an exaggerated respect, which can appear insulting, and a crass insensitivity, which is even more damaging. It comes as no surprise that cultural diversity has been identified as the single greatest barrier to business success.

It is no coincidence that the global explosion happened just as the e-commerce revolution arrived, with its 365 days a year/24 hours a day culture, allowing round the clock working and creating a market requiring international expertise backed by local knowledge and innovative management systems. Although it could be argued that in an e-commerce age cultural differences are likely to decrease in significance, they are in fact still very important, and remain major barriers to the globalisation of e-commerce. These differences extend also to commercial practice. Consider, for example, an issue that is crucial to e-commerce – privacy. The USA and Europe have traditionally had very different attitudes towards privacy, and Europe has adopted a very different approach to this problem vis-à-vis e-commerce. In the USA, the approach towards making e-commerce a secure environment has been one of industry self-regulation, whilst the EU has decided that legislation is more appropriate and effective, as manifested by the European Directives on e-commerce and electronic signatures. Even in a digital economy an organisation still needs to discover and analyse a client's values and preferences, and there is still a role for trading intermediaries such as banks, trading companies, international supply chain managers, chambers of commerce, etc. in helping to bridge differences in culture, language and commercial practice. In an era of global markets purists could perhaps say that splitting markets into European and global sectors is a contradiction. However, Europe does have its own unique features, not least its public procurement directives, physical link and proximity to the UK, and for some states a single currency. Therefore, this chapter will first consider the European market, before looking at other opportunities.

Europe

During the early 1990s Euromania broke out in the UK construction industry, and 1 January 1993 was to herald the dawn of new opportunity. It was the day the remaining physical, technical and trade barriers were removed across Europe, and from now on Europe and its markets lay at the UK construction industry's feet. Optimism was high within the UK construction industry – after all, it seemed as though a barrier free Europe with a multi-billion pound construction related output (€ 909 billion in 2004, according to the European Construction Industry Federation, see Table 7.1) was the solution to the falling turnover in the UK. Almost every month conferences were held on the theme of how to exploit construction industry opportunities in Europe. So, more than a decade later, has the promise been turned into a reality? There follows a review of developments in Europe, with an examination of the procurement opportunities for both the public and private sectors.

European public sector procurement – an overview

Procurement in the European public sector involves governments, utilities (i.e. entities operating in the water, energy and transport sectors) and local authorities purchasing goods, services and works over a wide range of market sectors, of which construction is a

Table 7.1 EU construction-related output – 2004 (Source: FIEC)

	Billion €	*Percentage*
Germany	208	22.9
Spain	131	14.4
United Kingdom	130	14.3
France	123	13.5
Italy	113	12.4
Netherlands	49	5.4
Belgium	23	2.5
Denmark	22	2.4
Ireland	22	2.4
Portugal	22	2.4
Austria	16	1.8
Sweden	17	1.9
Finland	19	2.1
Greece	14	1.5
Total	**909**	**100**

major part. For the purposes of legislation, public bodies are divided into three classes:

1. Central government and related bodies, e.g. NHS trusts
2. Other public bodies, e.g. local authorities, universities, etc.
3. Public utilities, e.g. water, electricity, gas, rail.

But why should quantity surveyors be interested? First, as a procurement professional the quantity surveyor has much to bring to the European procurement arena, not only for the existing member states but also for the new states that joined the EU club in 2005. Many EU states, and particularly but not exclusively those from the former Eastern Block countries, still employ procurement systems that favour the select few and preclude many, with all the consequences that this brings for value for money. Secondly, the widespread adoption of public private partnership throughout Europe as the preferred procurement route for a whole range of products, from the high profile TENs or Trans European Networks to the provision of local healthcare facilities, means that inevitably the quantity surveyor, whether in the public or private sector, will increasingly require a working knowledge of the rules governing the regulation of European public procurement. This is particularly so because there is increasing evidence that failure by a contracting authority to comply with the EU Directives will be severely punished, as was demonstrated in the so-called 'Harmon Case' (discussed later in this chapter), potentially leading to a multi-million pound price tag for damages. Finally, the sheer size and diverse nature of the market, both existing and enlarged, make Europe a real and exciting challenge.

The directives – theory and practice

The EU directives provide the legal framework for the matching of supply and demand in public procurement. A directive is an instruction addressed to the EU member states to achieve a given legislative result by a given deadline. This is usually done by transposing the terms of the directive into national legislation. Thus, the public works directives are instructions to the member states to modify their public procurement procedures to comply with the requirements of the directive. In France this was achieved by amending the *Code des Marchés Publics*, a code that is part of national law, to take account of EU legislation. In the UK, where there is no equivalent body of law governing public procurement, the government sought to achieve implementation by means of the Public Works Contracts Regulations 1991 and by issuing instructions in the form of circulars to all public purchasing authorities within the scope of the directives. Alarmingly, statistics from Directorate General XV (DGXV), the Brussels based directorate responsible for the regulation of the internal market, reveal that at any one time one in eight of all internal market directives had still to be implemented by member states.

The European public procurement regulatory framework was established by the public procurement Directives 93/36/EEC, 93/37/EEC and 92/50/EEC for supplies, works and services, and Directive 93/38/EEC for utilities, which, together with the general principles enshrined in the Treaty of Rome (1957), established the following principles for cross-border trading (references apply to the Treaty of Rome):

- A ban on any discrimination on the grounds of nationality (Article 6)
- A ban on quantitative restrictions on imports and all measures having equivalent effect (Articles 30 to 36)

- The freedom of nationals of one member state to establish themselves in another member state (Articles 52 et seq.) and to provide services in another member state (Articles 59 et seq.).

Enforcement Directives (89/665EEC and 92/13EEC) were added in 1991 in order to deal with breaches and infringements of the system by member states.

Although adopted in the 1990s the Directives date back to the 1970s, and almost immediately it became apparent that they failed to reflect the changes resulting from the information technology revolution and the liberalisation of telecommunications across Europe that is enabling the expansion of e-procurement (see Chapter 5). Public procurement is quite different from private business transactions in several aspects; the procedures and practices are heavily regulated and, whilst private organisations can spend their own budgets more or less as they wish (with the agreement of their shareholders), public authorities receive their budgets from taxpayers and therefore have a responsibility to obtain value for money, traditionally based on lowest economic cost. However, in recent years the clear blue water between private and public sectors has disappeared rapidly with the widespread adoption of public private partnerships and the privatisation of what were once publicly owned utilities or entities. The trend towards private involvement in public works has caused some difficulties within Directorate General XV, as for some time the legislation that empowers the system has been lagging behind developments in the market.

Unhindered cross-border tendering, free from tariffs, restrictions and protectionism, was the goal and the ongoing ideal of the European Commission and Directorate General XV in particular. The EU has legislated prolifically on public procurement since the 1985 White Paper Completing the Internal Market, which outlined the single market programme. The policy behind the creation of a single internal market was the belief that it could deliver greater economic performance and produce – according to the Cecchini report (1988), a 4.5–7 per cent increase in the Community GDP. The volumes in monetary terms of goods and services that are procured within the states of the EU are truly immense. Contracts for public works and for the purchase of goods and services by public authorities and utilities account for around 16.3 per cent of the Union's GDP according to the Internal Market Commission.

Given the historical background, it will come as no surprise that, more than 16 years after the introduction of European cross-border trading legislation, import penetration levels across all sectors are estimated to be only between 3 and 9 per cent. The statistics at the more optimistic end of the range are produced by DGXV, by taking into account motor car manufacture, which involves components produced in several EU states that then subsequently cross borders to be assembled in other member states. In addition, it is generally accepted that 85 per cent of public authorities do not comply with the directives in the knowledge that enforcement of compliance is virtually impossible. In 1996 the Commission published a Green Paper entitled *Public Procurement in the European Union: Exploring the Way Forward* as the basis for a dialogue about how to improve the system, in response to the following concerns of the member states:

- That member states were continuing to follow a buy national policy, in part because the directives were failing to produce a transparent and equal system
- That SMEs (small to medium-sized enterprises) were being excluded
- That the enforcement procedure to deal with breaches and infringements of the directives was totally inefficient (see below)
- That the legal framework was too complex and unsuitable for the 'electronic age'.

Following receipt of several hundred responses, in 1998 the Commission published Public Procurement in the European Union with proposals for reform, namely:

- Consolidation of the three classical sector directives previously described into a single directive for supply, works and services contracts
- A new directive for Utilities
- New rules on the use of competitive negotiated procedures to take account of public private partnerships
- New rules to permit and control framework purchasing (this will be described later in the chapter).

Popular opinion is that Brussels is bristling with civil servants on very large salaries who do very little in return. In reality, DGXV badly lacks the resources to operate an effective public procurement regime and consequently has to adopt a reactive rather than proactive approach, particularly to questions of enforcement.

Regrettably, the mechanisms for dealing with infringements of directives is to remain largely unchanged, and therefore it is almost inevitable that some member states will continue to take advantage of this. It is also to be regretted that the procurement directives were not drawn up by procurement professionals but rather by politicians and civil servants, whose first priority was the promotion of cross-border competition; consequently good procurement practice was not high on the agenda. Many procurement professionals see this fundamental flaw as the principal reason for the limited success of the directives. Another major concern (as mentioned previously) is the Commission's inability to enforce the directives and properly to punish contracting authorities who break the rules and cherry pick the parts of the law that they want to adhere to. Also regrettable is the lack of urgency shown by some member states in complying with directives and the manner in which compliance is achieved. For example, the Commission has instructed all member states to put in place a system for dealing with contracting authorities who break the public procurement rules and award contracts unfairly. The UK has chosen to interpret the directive by determining that the relevant forum for complaint within the UK shall be:

- the High Court in England and Wales
- the Court of Session in Scotland
- the High Court in Northern Ireland.

In practice this is referral to the existing legal system, with all the associated legal traps along the way. Whether this interpretation is in the spirit of the directives is debatable. Many think this is not an entirely appropriate method for dealing with disputes in what is considered to be a time sensitive environment; referral to these courts can take many months (or even years), by which time the project may well have been completed. Perhaps one of the most high-profile infringement cases was the award of the contract to construct the Stade de France, at Saint Denis in the outskirts of Paris, for the 1998 Football World Cup Finals. Not only does this case involve the granting of a concession, an area that will be dealt with later in this chapter, but it also illustrates the woefully inadequate mechanisms for rectifying breaches in the EU Directives. To understand the background to this case it is necessary to appreciate that in the mid-1990s France was in the midst of a deep recession, with unemployment levels of around 10 per cent and an impending presidential election. The construction of the Stade de France, like the World Cup Finals, was a high profile project, the non-completion of which would have had disastrous consequences for the tournament, as well as widespread political and financial implications for many organisations. During 1996 a formal complaint was made to the Commission concerning the contract award procedure for the stadium. However, through a combination of time wasting and public sector bureaucracy it was not until March 1998 – 2 months before the first match was played and the stadium completed – that the French Minister for European Affairs formally recognised the existence of the infringements. This admission of malpractice allowed a process that had been started nearly 2 years previously, and had included France being referred to the European Court of Justice by the then Single Market Commissioner Mario Monte, to be closed. However, no financial penalties were ever imposed, and the only outcome was a mild caution for the French contracting authority.

In a blatant foul, deserving of a red card, the French contracting authority permitted the winning French contractor to award a percentage of related construction contracts to local companies, in total breach of EU public procurement law. Equally importantly, the contract was described in the contract announcement as a concession rather than a works contract, although the final contract had all operating aspects removed; this effectively gave the wrong impression to prospective bidders and excluded many non-French contractors who were misled by the Official Journal announcement.

However, despite criticisms of the mechanisms by which complaints against contracting authorities who breach the directives are dealt with, recently, particularly in the UK, there have been a number of cases that have demonstrated for the first time that the UK remedies system does have teeth and that judges are prepared to take a robust attitude when it comes to dealing with breaches in the EU procurement law. Most notably, in October 1999 in the High Court, Judge Humphrey Lloyd QC delivered a decision in the case of Harmon CFEM Facades (UK) Ltd v The Corporate Officer of the House of Commons that attracted more interest than any procurement case so far before the UK courts. If this approach sets a precedent, contracting authorities have much more to fear from legal challenge than was thought to be the case. The case also has symbolic importance as the first major case in which an authority has been clearly condemned for a breach of the rules, and may now be open to significant damages liability as a result. The pity is that the judgement is nearly 300 pages long and took over a year to write – hardly a speedy and cost effective disputes resolution system! The case concerned a major £30 million contract for the fenestration work on the new office building for the House of Commons. The contract had been tendered under the Public Works Contract Regulations and the lowest bid, on a tender based on two options, was submitted by the French-based Harmon CFEM Facades (UK) Ltd in the sum of £29.56 million. However, the contract was awarded to the more expensive UK company Seele/Alvis, who submitted a tender price of £32.26 million. The reason for selecting Seele/Alvis was given by the House of Commons as the alleged 'commercial' nature of the bid. His Honour Judge Humphrey Lloyd was not convinced, and accused the Commons of adopting a buy British policy in breach of the European Public Procurement rules. An interim payment of £1.85 million for loss of profit was made in June 2000 and in the same year an out of court settlement was made of £5.2 million, including costs. Like a curate's egg, EU public procurement is not all bad practice and procrastination. By contrast to the system adopted in the UK for dealing with disputes and breaches, the Danish system is much faster. In 1992 Denmark established the Klagenoevnet for Udbud – The Complaints Board. The board is a quasi-judicial tribunal, the members being drawn from technical experts who are able to deal quickly with alleged breaches in public procurement law and dispense remedies and/or financial penalties where appropriate. The speed of referral to the board means that, unlike in France or the UK, contracts unfairly awarded can be declared void and new bids sought without jeopardising the project.

The quantity surveyor and EU public procurement

So what of the current system – how is the quantity surveyor likely to come into contact with the European public procurement juggernaut, and what are the potential pitfalls? The following scenarios will be illustrated:

- A surveyor working within a public body (contracting authority) and dealing with a works contract
- A surveyor in private practice wishing to bid for work in Europe as a result of a service contract announcement.

A surveyor within a public body

A quantity surveyor working within a body governed by public law (if in doubt, a list of European bodies and categories of bodies is listed in the directives) should be familiar with procedures for compliance with European public procurement law. The directives lay down thresholds above which it is mandatory to announce the contract particulars. The Official Journal is the required medium for contract announcements and is published five times each week, containing up to 1000 notices covering every imaginable contract required by central and local government and the utilities – from binoculars in Barcelona to project management in Porto. Major private sector companies also increasingly use the Official Journal for market research. The current thresholds (effective from January 2004) for announcements in the Official Journal are:

1. For works contracts (i.e. construction)   £3 834 411
2. For supplies and service contracts (i.e.
    quantity surveying, project management)
    • Central government   £99 695
    • Other public bodies   £153 376
3. Utilities: water, electricity, urban transport,
    airports, ports
    • Supplies and services   £306 753
    • Works   £3 834 411
    Oil, gas, coal and railways
    • Supplies and services   £258 923
    • Works   £3 236 542
    (NB: All figures exclude VAT).

Although, DGXV actively encourages contracting authorities and entities to announce contracts that are below threshold limits.

Information on impending tenders is published by the European Commission in the *Official Journal of the European Communities*, often otherwise known as the OJEU. While the Official Journal used to be produced as a printed publication it is now available on the internet. The official European Commission website for tenders, *Europa* can be accessed via europa.eu.int/, and provides free access to the information. There are a number of commercial services that provide access to the same information and often include additional details, such as *Tenders Direct* at www.tendersdirect.co.uk. Services such as Tenders Direct provide additional information and powerful

Search Results ————

Your search for '*Quantity Surveying*' found **164** documents published in the last 6 months. *Click* the reference of the document you wish to view. [Note: If you wish to search for older documents please use the advanced search facility.]

	Page 1 2 3 4 5 6		Next Page

Ref	Title	Published Deadline	
144110-2001	UK-Birmingham: project management, design, architectural, engineering, cost control and management s	31/10/01	29/11/01
144113-2001	UK-Glasgow: architectural, engineering, construction and related technical consultancy services	31/10/01	30/11/01
144133-2001	UK-London: building consultancy services	31/10/01	03/12/01
143454-2001	UK-Oxford: quantity surveying services	30/10/01	08/11/01
142416-2001	I-Milan: underground car park	27/10/01	17/12/01
141275-2001	UK-Stornoway: project-management services	25/10/01	19/11/01
141317-2001	UK-Oxford: architectural, engineering, construction and related technical consultancy services	25/10/01	07/11/01
140573-2001	UK-Epping: planning, design, project management and other related services	24/10/01	30/11/01
137785-2001	UK-Chelmsford: design services	18/10/01	09/11/01
137796-2001	IRL-Castlebar: design services	18/10/01	21/11/01
136991-2001	IRL-Dublin: architectural, engineering, construction and related technical consultancy services	17/10/01	
137147-2001	IRL-Tullamore: architectural, engineering and quantity surveying services	17/10/01	19/11/01
136442-2001	UK-Lewes: architectural, engineering and quantity surveying consultancy services	16/10/01	07/11/01
134999-2001	UK-Derby: advisory and information services	12/10/01	12/11/01
133625-2001	UK-Leeds: architectural, engineering and associated consultancy services	10/10/01	09/11/01
133653-2001	UK-Sheffield: quantity surveying and cost-management services	10/10/01	15/10/01
132022-2001	B-Brussels: topographical services	06/10/01	16/11/01
129304-2001	UK-Lancaster: construction-related professional-services	29/09/01	29/10/01
128448-2001	IRL-Dublin: building design team services	28/09/01	30/10/01
128456-2001	UK-Arbroath: architectural design, engineering and related services	28/09/01	09/10/01
127734-2001	IRL-Tullamore: architectural, engineering and quantity surveying services	27/09/01	
127821-2001	UK-Ripley: building-consultancy services	27/09/01	26/10/01
127858-2001	IRL-Castlebar: architectural design and associated services	27/09/01	31/10/01
127870-2001	IRL-Dublin: architectural, engineering and quantity surveying services	27/09/01	16/11/01
127161-2001	UK-Birmingham: architectural design, engineering, quantity surveying and related-services	26/09/01	23/10/01
127168-2001	UK-Wolverhampton: architectural, civil engineering and related works and services	26/09/01	29/10/01
126563-2001	IRL-Castlebar: design services	25/09/01	31/10/01
125022-2001	UK-Worcester: architectural, engineering, project management and related services	22/09/01	22/10/01
124602-2001	UK-Shrewsbury: professional design and associated services	20/09/01	03/10/01
123216-2001	UK-Arbroath: architectural design, engineering and related services	18/09/01	

	Page 1 2 3 4 5 6	Next Page

Figure 7.3 Tenders Direct results.

easy-to-use search facilities that enable relevant tenders to be identified, as well as an e-mail alert service to provide notification of relevant tenders in the future (see Figure 7.3).

Figure 7.3 shows the search results screen following a search for quantity surveying contracts. All the notices that match the search criteria are displayed in chronological order, with the most recent at the top of the list. The list includes the location and title of the project as well as the date of publication and, crucially, the deadline by which a supplier must have confirmed their interest in tendering. An abstract of the notice can be viewed by clicking on its reference number, although to obtain the full tender notice users are required to register with Tenders Direct and pay a small fee for each notice they wish to download. It should be noted that the search results also show UK public procurement opportunities for quantity surveyors, and as such can be useful to UK-based organisations too.

As mentioned previously, the EU public procurement directives have been around for about 30 years; however, in 1996 a process of updating was commenced in order to take account of the many changes that had taken place during the last few years. In 2004 the changes were agreed and adopted by the European Parliament and the changes must be transposed into national legislation by 31 January 2006. Significantly, both the Scottish Executive and the DTI have warned that they will not be able to implement the new EU legislation by the deadline and that there is likely to be a gap of a few months between when the new directives should have been implemented and the date when they are actually implemented in the UK. The new directives, it is hoped, will provide a framework within which public procurement must be conducted and have been widely drawn to cater for all 25 member states. The OGC launched a public consultation looking for views on how the Consolidated Procurement Directive can be best implemented.

An interesting development has been the decision by the Scottish Executive that it wishes to have its own set of Scottish implementing regulations that will be distinct from those in England and Wales. There has not yet been any convincing justification given for this decision, but it has certainly proved universally unpopular, judging by the recent responses to the recent public procurement consultation. Concerns have been expressed by contracting authorities and utilities that this will simply lead to increased compliance costs, as procurement lawyers will need to become familiar with two sets of implementing regulations and with any differences between the two. Additionally, there is concern that ambiguity could be caused by the situation where, for example, a Scottish contracting authority runs a procurement exercise for services to be delivered south of the border or there is a need for services to be delivered on both sides of the border. There is doubt as to which set of rules would apply or whether both sets will.

The consolidated Procurement Directive does not seek to bring about radical change, rather its aim is to modernise the existing directives by clearly permitting electronic means of procurement and to facilitate new procurement models, such as PPPs. The directive also clarifies existing law in areas such as the selection of tenderers and the award of contracts, bringing the law as stated into line with judgements of the European Court of Justice.

The changes to the procurement process can be summarised as follows:

1. A consolidated single directive: one for the public sector and one for the utilities to replace the existing one
2. Refinement of existing procurement provisions to include:
   - Simplified thresholds expressed in Euros, available from the OGC website from January 2006 – see previous reference
   - Encouragement to use performance specifications, in all forms of procurement, not just PPPs
   - Environmental and social issues addressed
   - The expansion of electronic tendering (see Chapter 5) and communication which can be utilised to reduce the bid period by 7 days compared with conventional means.
3. Significant additions include provisions that impact on PPPs such as:
   - The Competitive Dialogue procedure, recommended for use with PFI projects and
   - Framework agreements used for ProCure 21 and Partnerships for Schools.

The announcement procedure involves three stages:

1. Prior information notices (PIN) or indicative notices
2. Contract notices
3. Contract award notices (CANs).

Examples of these notices can be found in Annex IV of the Directive, and examples of all three types of notice are given in the Appendix to this book.

A prior information notice, or PIN, that is not mandatory, is an indication of the essential characteristics of a works contract and the estimated value. It should be confined to a brief statement, and posted as soon as planning permission has been granted. The aim is to enable contractors to schedule their work better and allow contractors from other member states the time to compete on an equal footing. Where work is subdivided into several lots, each on the subject of a contract, the aggregate value should be taken into account when determining whether the threshold has been exceeded. For example, a contract for the construction of a new prison has been divided into three lots, estimated at € 3 million, € 2.2 million and € 1.6 million, respectively. The estimated value for procurement processes is therefore € 6.8 million, and the directives will apply. It is a common complaint that authorities split contracts to avoid the directives; although specifically prohibited by the directives this action is difficult to prove, but it does help to promote a buy national policy by states. The prior information notice can be a useful market-testing tool in the case of public private partnerships, as it affords the contracting authority the opportunity to assess the potential interest from consortia, as well as the financial viability and business case of the project that is being proposed. Increasingly the commission is encouraging public bodies to post a purchaser profile, which is a statement of the kind of supplies, services and works contracts that a particular body is likely to require, and at what intervals and quantities.

Contract notices are mandatory and must include the award criteria, which can be based on either the lowest price or the most economically advantageous tender, specifying the factors that will be taken into consideration.

Once drafted, the notices are published, five times a week, via the Publications Office of the European Commission in Luxembourg in the Official Journal via the Tenders Electronic Daily (TED) database, and translated into the official languages of the community, all costs being borne by the community. TED is updated twice weekly and may be accessed through the Commission's website at http://simap.eu.int. Extracts from TED are also published weekly in the trade press.

In order to give all potential contractors a chance to tender for a contract, the directives lay down minimum periods of time to be allowed at various stages of the procedure – for example, in the case of Open Procedure this ranges from 36 to 52 days from the date of dispatch of the notice for publication in the Official Journal. Restricted and Negotiated procedures have their own time limits. These timescales should be greatly reduced with the wide-scale adoption of electronic procurement (see Figure 7.4).

The production of tender documents for a construction project for the domestic market is in itself a difficult task that requires a good deal of experience and professional know how. To produce documentation suitable for transmission to all EU member states in a form that maintains transparency and accuracy demands due diligence of the highest order. To aid contracting authorities in accurately describing proposed works, the Common Procurement Vocabulary (CPV) has been developed to facilitate fast and accurate translation of contract notices for publication in the Official Journal. CPV is a series of nine digit codes that relate to the area of goods, services or works for which tenders are invited; at present the use of CPV in contract notices is optional. Certainly CPV makes searching for particular market sectors on databases much easier than has previously been the case. Although the use of CPV is currently not mandatory, if the contracting entity does not use the classification system the Commission will translate it, with perhaps unusual if not entirely unpredictable results. For example, the term 'Christmas trees' is widely used within the oil and gas industries as vernacular for blow-out valves. Consequently, when Christmas trees were referred to by an entity in oil and gas but not classified with CPV in the tender notice, the commission assumed that the entity was attempting to procure trees and classified the notice accordingly! However, in August 2001 the European Commission adopted a proposal that will establish CPV as the only system used for the classification of public procurement announcements. The motive behind the decision was to ensure that the subject matter of contracts could be accurately identified, allowing automatic translation of tender notices into all official Community languages.

Contract award notices inform contractors about the outcome of the procedure. If the lowest price was the standard criterion, this is not difficult to apply. If, however, the award was based on the 'most economically advantageous tender', then further clarification is required to explain the criteria – e.g. price, period for completion, running costs, profitability and technical merit, listed in descending order of importance. Once established, the criteria should be stated in the contract notices or contract documents.

Award procedures

The surveyor must decide at an early stage which award procedures should be adopted.The new version of the directives make the following procedures available:

1. Open procedure
2. Restricted procedure
3. Competitive dialogue
4. Negotiated procedure.

The following pages will concentrate on those procedures most commonly used in PPP models in the UK. Generally in order to comply with the procurement directives the following rules should be followed:

- Minimum number of bidders must be five for the restricted procedure and three for the negotiated and competitive dialogue procedures
- Contract award is made on the basis of lowest price or most economically advantageous tender (MEAT)
- Contract notices or contract documents must provide the relative weighting given to each criterion used to judge the most economically advantageous tender and where this is not possible, award criteria must be stated in descending order of importance
- MEAT award criteria may now include environmental characteristics, for example, energy savings, disposal costs, provided these are linked to the subject matter of the contract.

Competitive dialogue

Article 1 (11c) defines Competitive Dialogue as follows:

A procedure in which any economic operator may request to participate and whereby the contracting authority conducts a dialogue with the candidates admitted to that procedure, with the aim of developing one or more suitable alternatives capable of meeting its requirements and on the basis of which the candidates chosen will be invited to tender.

Article 29(1) describes its use

For particularly complex contracts where use of the open or restricted procedures will not allow the award of the contract.

The introduction of this procedure addresses the need to grant, in the opinion of the European Commission, contracting authorities more flexibility to negotiate on PPP projects. Some contracting authorities have complained that the existing procurement rules are too inflexible to allow a fully effective tendering process. Undoubtedly, the degree of concern has depended largely on how a contracting authority has interpreted the procurement rules as there are numerous examples of PPP/PFI projects which have been successfully tendered since the introduction of the public procurement rules using the

Negotiated Procedure. However, the European Commission recognised the concerns being expressed, not only in the UK but also across Europe, and it has sought to introduce a new procedure which will accommodate these concerns. In essence, the new competitive dialogue procedure permits a contracting authority to discuss bidders' proposed solutions with them before preparing revised specifications for the project and going out to bidders asking for modified or upgraded solutions. This process can be undertaken repeatedly until the authority is satisfied with the specifications that have been developed. Some contracting authorities are pleased that there is to be more flexibility in negotiations; however, for bidders this reform does undoubtedly mean that tendering processes could become longer and more complex. This in turn would lead to more expense for bidders and could pose a threat to new entrants to the PPP market, as well as existing players. According to the Commission's DGXV department, the introduction of this procedure will enable:

- Dialogue with selected suppliers to identify and define solutions to meet the needs of the procuring body
- Awards to be made only on the basis of the most economically advantageous basis.

In addition,

- All candidates and tenderers must be treated equally and commercial confidentiality must be maintained unless the candidate agrees that information may be passed on to others
- Dialogue may be conducted in successive stages. Those unable to meet the need or provide value for money, as measured against the published award criteria, may drop out or be dropped, although this must be conveyed to all tenderers at the outset
- Final tenders are invited from those remaining on the basis of the identified solution or solutions
- Clarification of bids can occur pre- and post-assessment provided this does not distort competition.

To summarise therefore, the Competitive Dialogue Procedure is, according to the commission, to be used in cases where it is difficult to access what would be the best technical, legal or financial solution because of the market for such a scheme or the project being particularly complex. However, the Competitive Dialogue Procedure leaves many practical questions over its implementation, for example:

- The exceptional nature of the Competitive Dialogue and its hierarchy with other award procedures
- The discretion of the contracting authorities to initiate the procedure, who is to determine the nature of a particular complex project
- The response of the private sector, with particular reference to the high bid costs
- The overall value for money
- The degree of competition achieved as there is great potential for post-contract negotiations.

Compared with Negotiated Procedure, Competitive Dialogue is up to 22 days longer.

Until now the Negotiated Procedure has been used by UK public entities when procuring a PFI project and this will still be available to project managers and the devil appears to be in the detail, particularly with defining the 'particularly complex' criteria. The differences between the negotiation and competitive dialogue procedures are illustrated in Table 7.2.

Framework agreements

Article 1(5) defines a framework as:

> *An agreement between one or more contracting authorities and one or more economic operators, the purpose of which is to establish the terms governing contracts to be awarded during a given period, in particular with regard to price and where appropriate, the quantity envisaged.*

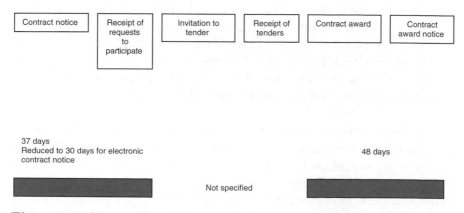

Figure 7.4 Competitive Dialogue, minimum timescales.

Table 7.2 Competitive Dialogue and Negotiated Procedure compared

Competitive Dialogue	Negotiated Procedure
Always involves competitive tender and can only use the most economically advantageous award criteria (MEAT)	Competitive tender not required. Can use MEAT or lowest price and in specified and limited circumstances, can negotiate with a single supplier
Dialogue may embrace all aspects of the contract for the purpose of identifying one or more solutions which meet the purchaser's needs before seeking bids from those remaining in the dialogue	Negotiation following advertisement is intended to adapt tenders received in order to better meet the purchaser's specific needs
	Used to allow negotiation when: • Competition is not viable or appropriate • Other procedures have not produced an acceptable tender • Works are needed for R&D purposes • Prior overall pricing is not possible • Services cannot be specified sufficiently precisely for use of open or restricted procedures

In other words, a framework establishes the terms and conditions that will apply to subsequent contracts (call offs) but does not create right and obligation. Frameworks can cover supplies, works and services and can be used in conjunction with the Open, Restricted, Competitive dialogue and Negotiated Procedures. A framework agreement is used in the NHS ProCure 21 procurement strategy described in Chapter 4.

The position regarding the legality of framework agreements under the existing public sector directives was unclear; although use of these procurement models was not specifically provided for, neither was it prohibited. As a result, much debate sprang up regarding the legality of framework agreements as a method of procurement by bodies bound by public sector rules. The Consolidated Procurement Directive clarifies this position, specifically regulating the use of a single supplier and multi-supplier models and the call-off of individual contracts under the framework itself. Where there is only one supplier signed up to a framework agreement, contracts under it must be awarded by applying the rules set out in the framework agreement, although the contracting authority may ask the supplier to supplement its tender where necessary. In the case of a multi-supplier framework agreement at least three suppliers must be signed up to it and depending on whether all the terms are set out in the framework agreement, a contracting authority may award contracts by applying those terms without re-opening competition, or it may conduct a mini-competition. The directive sets out how such a mini-competition should be carried out.

The Consolidated Procurement Directive also regulates the duration of framework agreements stating that they may not be longer than 4 years unless there are duly justified and exceptional circumstances. What constitutes 'exceptional circumstances' is not defined; however, it is likely to encompass situations where a longer period is justified by reason, for example, of a contractor's investment in the contract. In such cases the duration should be sufficiently long to permit recuperation of that investment plus a reasonable amount of profit.

The EU directive outlines the principals that must be satisfied before a framework can operate as shown in Figure 7.5.

Finally, as PPP contracts are generally long term, a new provision to base the contract award criteria on environmental characteristics provided that these are linked to the subject matter of the contract and looked at from the point of view of the contracting authority, e.g. running costs, energy costs and additional environmental quality (toxic emissions), etc.

Electronic auctions

The internet is making the use of electronic auctions increasingly more attractive as a means of obtaining bids in both public and private sectors; indeed it can be one of the most transparent methods of procurement. At present electronic auctions can be used in both Open and Restricted framework procedures. The system works as follows:

- The framework (i.e. of the selected bidders) is drawn up
- The specification is prepared
- The public entity then establishes the lowest price award criterion, for example with a benchmark price as a starting point for bidding

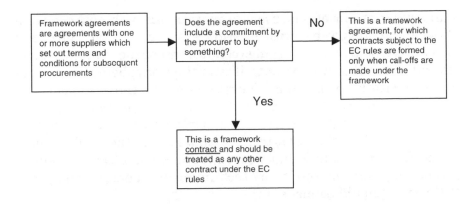

The call-off stage

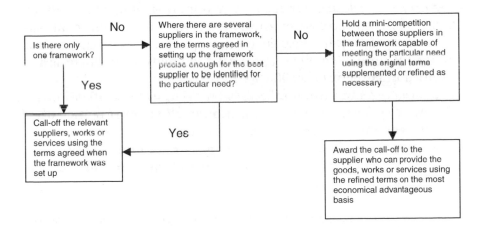

Figure 7.5 Framework agreements.

- Reverse bidding on a price then takes place, with framework organisations agreeing to bid openly against the benchmark price
- Prices/bids are posted up to a stated deadline
- All bidders see the final price.

Technical specifications

At the heart of all domestic procurement practice is compliance with the technical requirements of the contract documentation in order to produce a completed project that performs to the standards of the brief. The project must comply with national standards to be compatible with existing systems and technical performance. The task of achieving technical excellence becomes more difficult when there is the possibility of the works being carried out by a contractor who is unfamiliar with domestic conventions and is attempting to translate complex data into another language. It is therefore very important that standards and technical requirements are described in clear terms with regard to the levels of quality, performance, safety, dimensions, testing, marking or labelling, inspection, and methods or techniques of construction, etc. References should be made to:

- A *Standard*: a technical specification approved by recognised standardising body for repeated and continuous application
- A *European Standard*: a standard approved by the European Committee for Standardisation (CEN)
- *European technical approval*: a favourable technical assessment of the fitness for use of a product, issued by an approval body designated for the purpose (sector-specific information regarding European technical approval for building products is provided in Directive 89/106/EEC)
- *Common technical specification*: a technical specification laid down to ensure uniform application in all member states, which has been published in the Official Journal
- *Essential requirements*: requirements regarding safety, health and certain other aspects in the general interests that the construction works must meet.

Given the increased complexity of construction projects, the dissemination of accurate and comprehensive technical data is gaining in importance. It is therefore not surprising that the Commission is concerned that contracting authorities are, either deliberately or otherwise, including discriminatory requirements in contract documents. These include:

- Lack of reference to European standards
- Application of technical specifications that give preference to domestic production
- Requirements of tests and certification by a domestic laboratory.

The result of this is in direct contravention of Article 30 of the Treaty of Rome and effectively restricts competition to domestic contractors. In an attempt to reduce the potential problems outlined above, the EU has embarked on a campaign to encourage contracts to be based on an output or performance specification, which removes the need for detailed and prescriptive documentation.

A surveyor in private practice

Professional surveyors cannot help but be tempted by the calls for bids to be submitted to provide a whole range of services to contracting authorities through Europe. This section of the chapter should not be read in isolation; the preceding section should first be studied in order to appreciate fully the general public procurement environment. Anecdotal information of unsuccessful and costly attempts to break into the European public procurement market abounds; many of the difficulties are attributable to lack of knowledge of the directives and basic language competencies, as well as a large degree of naivety. It should be remembered that the European market is not a level playing field and that any attempt to break into new markets requires detailed market research, including the establishment of local contacts, for in this respect there is no difference between public and private sectors. When selecting member states to target, research must be carried out to discover whether the directives have been implemented – after all, there is little point in trying to win work in a member state that cherry-picks which parts of EU law it implements. In a review concluded in July 2001 by DGXV, the best performing countries were Denmark, Finland, Sweden and Spain, while Greece, France and Belgium were lagging behind with implementation deficits of more than twice those of the best performing countries. Attitude to dealing with infringements is also a good indicator with Denmark once more faring well compared with France, Germany and the UK. Other positive indicators are the procedures that are in place for dealing with infringements – for example, as in Denmark is there a separate national review board for dealing quickly with complaints.

Following initial selection, there are other checks that the surveyor should make when carrying out a search of contract announcements:

- When searching TED for PIN or contract notices, look for announcements with open award procedures and a comprehensive specification; this is a good indication that the contracting authority is determined, at least on the face of it, to make the procurement process as transparent as possible.

- Avoid announcements with single or multi-named contacts, it's far better to have a general administrative person as a contact for further information. Single name points of contact could indicate that the named person is carrying out an unofficial pre-selection screening!
- Avoid announcements with lists of acceptable tenderers. Although widely condemned by the DGXV this still exists and it is obviously a restrictive practice.
- Check the Contract Award Notices to discover which member states seem to be awarding a high proportion of contracts to domestic contractors. This task can be carried out quite simply via TED.
- Check the countries that use negotiated or accelerated procedures to excess, as this procurement path effectively excludes all but domestic contractors.

What therefore is the most effective way for a surveyor to access information concerning European public procurement opportunities?

As described previously, there are several pathways to access information on current contracts, ranging from freely available services to subscription services such as Tenders Direct, and logging on to these services could be a good starting point. Familiarity with the Common Procurement Vocabulary, an example of which is given in Table 7.3, will save time when carrying out a search, particularly if TED is being used. The appropriate CPV inserted into code in the search engine of TED will retrieve all the information on the database associated with a particular market sector. This facility may also be used to analyse tender award notices.

Table 7.3 Example of common procurement vocabulary

74000000	Architectural, construction, legal, accountancy and business services
	74200000 Architectural, engineering, construction and related technical consultancy services
	74210000 Technical consultancy services
	7426000 Construction-related services

It is a fact of life that the larger the organisation, the greater the resources, both physical and financial, that are available to take advantage of new markets and opportunities. One of the major concerns that emerged from the public procurement green paper was the way in which SMEs were being disadvantaged, a concern shared by DGVX, which, by a number of initiatives, has attempted to redress the balance of opportunity. One of the primary means of redressing the balance is the interface between procurement and information technologies. The Commission is promoting, through a series of directives and other initiatives, the internet as the preferred method of public procurement. The European Commission-funded SIMAP project was launched in order to encourage best practice in the use of information technology for public procurement. The development of information technologies and the dramatic reduction in telecommunication costs have created favourable conditions for moving from paper based commerce towards full e-commerce and addressing the whole procurement process, including bids, awards of contracts, delivery and invoicing.

The primary objectives of SIMAP are to improve the dissemination and quality of information on procurement opportunities and encourage electronic data interchange between purchasers and suppliers. It is thought that an electronic procurement system will reduce transaction costs and time, and improve the management of the system as a whole. The development of electronic procurement is a joint responsibility between the public and private sectors; the public authorities' role is to define the legal framework and to facilitate development of ICT tools, while the private sector must develop and implement applications and ensure technical interoperability.

Public procurement beyond Europe

There are no multi-lateral rules governing public procurement. As a result, governments are able to maintain procurement policies and practices that are trade distortive. That many governments wish to do so is understandable; government purchasing is used by many as a means of pursuing important policy objectives that have little to do with economics – social and industrial policy objectives rank high amongst these. The plurilateral Government Procurement Agreement (GPA) partially fills the void. GPA is based on the GATT provisions negotiated during the 1970s, and is reviewed and refined at meetings (or rounds) by ministers at regular intervals. Its main objective is to open up international procurement markets by applying the obligations of non-discrimination and transparency to the tendering procedures of government entities. It has been estimated that market opportunities for public procurement increased 10-fold as a result of the GPA. The GPA's approach follows that of the European rules. The agreement establishes a set of rules governing the procurement activities of member countries and provides for market access opportunities. It contains general provisions prohibiting discrimination as well as detailed award procedures. These are quite similar to those under European regime, covering both works and other services involving, for example, competition, the use of formal tendering and enforcement, although the procedures are generally more flexible than under the European rules. However, GPA does have a number of shortcomings. First, and perhaps most significantly, its disciplines apply only to those World Trade Organisation members that have signed it. The net result is a continuing black hole in multilateral WTO rules that denies access or provides no legal guarantees of access to billions of dollars of market opportunities in both the goods and services sector. The present parties are the European Union, Aruba, Norway, Canada, Israel, Japan, Liechtenstein, South Korea, the USA, Switzerland and Singapore.

Developments in public procurement

As in the private sector, information technology is the driving force in bringing efficiency and added value to procurement. However, despite the many independent research projects that have been undertaken by the private sector, the findings cannot simply be lifted and incorporated into the public sector due to the numerous UK and European Community regulations that must be adhered to. Notwithstanding these potential problems, the UK Government has set an ambitious target for the adoption of e-tendering in the public sector.

Of all strands of the e-business revolution, it is e-procurement that has been the most broadly adopted, has laid claim to the greatest benefits and accounts for the vast majority of electronic trading. A survey carried out on behalf of the EU in 2000 showed that, of the existing electronic procurement systems in use, building and construction was offered by all of them and was the top ranked sector, with a usage rate of 72 per cent.

Europe and beyond

The effect of culture on surveyors operating in international markets

As discussed in the opening of this chapter, culture can be a major barrier to international success. Culture must first be defined and then analysed so that it can be managed effectively; thereafter, there is the possibility of modelling the variables as an aid to business. A business culture does not change quickly, but the business environment from which it is derived and with which it constantly interacts is sometimes subject to radical and dramatic change. The business culture in a particular country grows partly out of what could be called the current business environment of that country. Yet business culture is a much broader concept, because alongside the impulses that are derived from the present business environment there are historical examples of the business community. For example, as discussed in Chapter 1, the 1990 recession saw widespread hardship, particularly in the UK construction industry. There have been many forecasts of doom during the early twenty-first century from analysts drawing comparisons between the state of business at this time with that in 1990, when record output, rising prices and full employment were threatening to overheat the economy as well as construction – can there be many quantity surveying practices in the UK that are not looking over their shoulders to see if and when the next recession is coming? Table 7.4 outlines a sample of the responses by 1500 European companies questioned during a study into the effects of culture on business.

So what is culture? Of the many definitions of culture, the one that seems most accurately to sum up this complex topic is 'an historical emergent set of values'. The cultural differences within the property/construction sectors can be seen to operate at a number of levels, but can be categorised as follows:

1. Business/economic factors – e.g. differences in the economic and legal systems, labour markets, professional institutions, etc., of different countries.
2. Anthropological factors, as explored by Hofstede (1984). The Hofstede IBM study involved 116000 employees in 40 different countries, and is widely accepted as being the benchmark study in this field.

Table 7.4 Effects of culture on business

China	Cultural differences are as important as an understanding of Asian or indeed other foreign languages
Far East	One needs to know etiquette/hierarchical structure/manner of conduct in meetings
Germany	Rigid approach to most operational procedures
Middle East	Totally different culture – time, motivation, responsibility
Russia	Inability to believe terms and conditions as stated really are what they are stated to be
SE Asia	Strict etiquette of business in South Korea and China can be a major problem if not understood
France	Misunderstandings occurred through misinterpretation of cultural differences

Of these two groups of factors, the first can be regarded as fairly mechanistic in nature, and the learning curve for most organisations can be comparatively steep. For example, the practice of quantity surveyors in France of paying the contractor a sum of money in advance of any works on site may seem risky, but it is usual practice in a system where the contractor is a trusted member of the project team. It is the second category of cultural factors, the anthropological factors, that is more problematic. This is particularly so for small and medium enterprises, as larger organisations have sufficient experience (albeit via a local subsidiary) to navigate a path through the cultural maze.

Perhaps one of the most famous pieces of research on the effects of culture was carried out by Gert Hofstede for IBM. Hofstede identified four key value dimensions on which national culture differed (Figure 7.6), a fifth being identified and added by Bond in 1988 (Hofstede and Bond, 1988). These value dimensions were power difference, uncertainty avoidance, individualism/collectivism, and masculinity/femininity, plus the added long-/short-termism. Although neatly categorised and explained in Figure 7.6, these values do of course in practice interweave and interact to varying degrees.

- *Power distance* indicates the extent to which a society accepts the unequal distribution of power in institutions and organisations, as characterised by organisations with high levels of hierarchy, supervisory control and centralised decision making. For example, managers in Latin countries expect their position

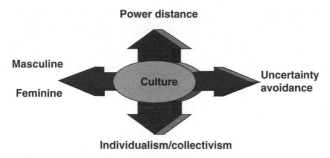

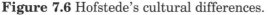

Figure 7.6 Hofstede's cultural differences.

within the organisation to be revered and respected. For French managers, the most important function is control, which is derived from hierarchy.

- *Uncertainty avoidance* refers to a society's ability to cope with unpredictability. Managers avoid taking risks and tend to have more of a role in planning and co-ordination. There is a tendency towards a greater quantity of written procedures and codes of conduct. In Germany, managers tend to be specialists and stay longer in one job, and feel uncomfortable with any divergence between written procedures – for example, the specification for concrete work and the works on site. They expect instructions to be carried out to the letter.
- *Individualism/collectivism* reflects the extent to which the members of a society prefer to take care of themselves and their immediate families as opposed to being dependent on groups or other collectives. In these societies, decisions would be taken by groups rather than individuals, and the role of the manager is as a facilitator of the team (e.g. Asian countries). In Japan, tasks are assigned to groups rather than individuals, creating stronger links between individuals and the company.
- *Masculinity/femininity* refers to the bias towards an assertive, competitive, materialistic society (masculine) or the feminine values of nurturing and relationships. Masculine cultures are characterised by a management style that reflects the importance of producing profits, whereas in a feminine culture the role of the manager is to safeguard the well being of the workforce. To the American manager, a low head count is an essential part of business success and high profit; anyone thought to be surplus to requirements will be told to clear his/her desk and leave the company.

As a starting point for an organisation considering looking outside the UK for work, Figure 7.7 may be a somewhat light-hearted but useful discussion aid to help recognise and identify the different approaches to be found towards organisational behaviour in other

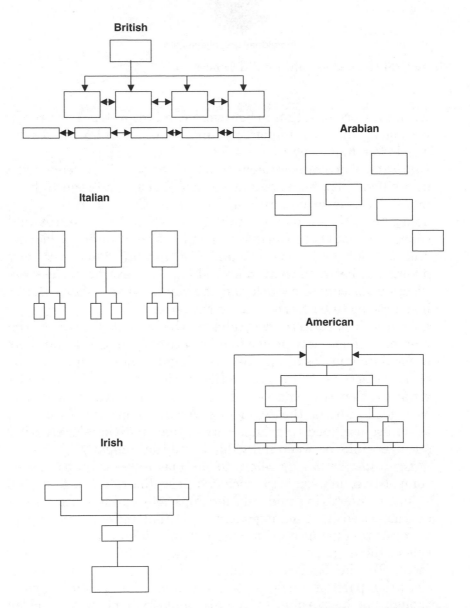

Figure 7.7 Organisational chart (Source: Adapted from International Management, Reed Business Publishing).

countries/cultures – approaches that if not recognised can be a major roadblock to success.

Developing a strategy

The development process, when carried out internationally, is particularly complex to manage due to the weaving together of various cultures, including language (both generic and technical), professional standards and construction codes, design approaches and technology, codes of conduct, and ethical standards. Technical competency and cultural integration must be taken as read.

The competencies necessary to achieve cultural fluency can therefore be said to be:

- Interpersonal skills
- Linguistic ability
- Motivation to work abroad
- Tolerance of uncertainty
- Flexibility
- Respect
- Cultural empathy.

Case studies of SMEs show that 60 per cent of companies react to an approach from a company in another country to become involved in international working. The advantages of reacting to an enquiry are that this approach involves the minimum amount of risk and requires no investment in market research, but consequentially it never approaches the status of a core activity, it is usually confined to occasional involvement and is only ever of superficial interest. However, to be successful the move into overseas markets requires commitment, investment and a good business plan linked to the core business of the organisation.

A traditional approach taken by many surveying practices operating in world markets, particularly where English is not the first language, is to take the view that the operation should be headed up by a native professional, based on the maxim that, for example, 'it takes an Italian to negotiate with an Italian'. Although recognising the importance of cultural diversity, the disadvantages of this approach are that the parent company can sometimes feel like a wallflower, there is no opportunity for parent company employees to build up management skills, and in the course of time local professionals may decide to start their own business and take the local client base with them. If culture is defined as shared values and beliefs, then no wonder so many UK companies take this approach. How long, for example, would it take for a British quantity surveyor to acquire the cultural values of Spain?

As a starting point, a practice considering expanding into new markets outside the UK should undertake the following:

1. Carry out extensive market research:
    *   Ensure market research covers communication (language and cultural issues)
    *   Make frequent visits to the market; it shows commitment rather than trying to pick up the occasional piece of work
    *   Use written language to explain issues, since verbal skills may be less apparent
    *   Use exhibitions to obtain local market intelligence and feedback
2. Ensure documentation is culturally adapted and not literally translated:
    *   Brochures should be fully translated into local language on advice from local contacts
    *   Publish new catalogues in the local culture
    *   Set up the website in the local language, with the web manager able to respond to any leads – after all, if a prospective client is expecting a fast response, waiting for a translator to arrive is not the way to provide it
    *   Adapt the titles of the services offered to match local perceptions
    *   Emphasise added value services
3. Depending on the country or countries being targeted, operate as, for example, a European or Asian company, rather than a British company with a multilingual approach – think global, act local:
    *   Arrange a comprehensive, multi-level programme of visits to the country
    *   Set up a local subsidiary company or local office or, failing that, set up a foreign desk inside the head office operating as if it is in the foreign country (keeping foreign hours, speaking foreign language, etc.)
    *   Change the culture of the whole company at all levels from British to European, Asian, etc. as relevant

- Recruit local agents that have been educated in the UK, so they have a good understanding of UK culture too
4. Implement a whole company development strategy:
   - Language strategy should be an integral part of a company's overall strategy as a learning organisation
   - Identify the few individuals who can learn languages quickly and build on this
   - Create in-house language provision
   - Set up short-term student placements in the UK for foreign students, via a sponsored scheme such as the EU Erasmus programmes
   - Target markets whose specialist language ability gives a competitive edge, e.g. China
5. Subcontract the whole export process to a specialist company:
   - Hire a company to provide an export package of contacts, liaison, translation, language training, etc.
6. Pool resources with other companies:
   - Share language expertise and expenses with other companies
7. In joint ventures, collaboration can be based on:
   - Equity/operating joint ventures, in which a new entity is created to carry out a specific activity. Seen as a long-term commitment, the new entity has separate legal standing
   - Contractual ventures, in which no separate entity is created and instead firms co-operate and share the risk and rewards in clearly specified and predetermined ways. On the face of it, this form of joint venture appears to be more formal
8. Management contracts:
   - The transfer of managerial skills and expertise in the operation of a business in return for renumeration.

Conclusion

With the advent of electronic communications, the possibilities that exist for quantity surveyors to operate on a European or global level have never been greater or easier to access. However, despite what some multinational organisations would have us believe, the world is not a bland homogeneous mass and organisations still need to pay attention to the basics of how to conduct interpersonal relationships if they are to succeed.

Bibliography

Bardouil, S. (2001). Surveying takes on Europe. *Chartered Surveyor Monthly*, Jul/Aug, pp. 20–22.

Brooke, M.Z. (1996). *International Management*, 3rd edn. Stanley Thornes.

Button, R. and Mills, R. (2000). Public sector procurement: The Harmon case. *Chartered Surveyor Monthly,* May 24.

Cartlidge, D. (1997). It's time to tackle cheating in EU public procurement. *Chartered Surveyor Monthly*, November/December, pp. 44–45.

Cartlidge, D. and Gray, C. (1996). *Cross Border Trading for Public Sector Building Work within the EU*. European Procurement Group, Robert Gordon University.

Cecchini, P. (1988). *The European Challenge*. Wildwood House.

Commission of the European Communities (1985). *Completing the Internal Market*. Office for Official Publications of the European Communities.

Commission of the European Communities (1994). *Public Procurement in Europe: The Directives*. Office for Official Publications of the European Communities.

Commission of the European Communities (1998). *Single Market News – The Newsletter of the Internal Market DG*. Office for Official Publications of the European Communities.

Hagen, S. (1998). *Business Communication Across Borders: A Study of the Language Use and Practice in European Companies*. The Centre for Information on Language Teaching and Research.

Hagen, S. (ed.) (1997). *Successful Cross-Cultural Communication Strategies in European Business*. Elucidate.

Hall, E.T. (1990). *Understanding Cultural Differences*. Intercultural Press.

Hall, M.A. and Jaggar, D.M. (1997). Should construction enterprises work internationally, take account of differences in culture? Proceedings of the Thirteenth Annual ARCOM 97 Conference, King's College Cambridge, 15–17 September.

Hofstede, G. (1984). *Culture Consequences: International Differences in Work-Related Values*. Sage Publications.

Hofstede, G. and Bond, M.H. (1988). *The Confuscius Connection: From Cultural Roots to Economic Growth*. Organizational Dynamics.

McKendrick, P. (1998). RICS Annual Report and Accounts: President's statement. *Chartered Surveyor Monthly*, February 2.

Tisser, M. *et al.* (1996). Chartered surveyors: an international future. *Chartered Surveyor Monthly,* October 15.

Websites

http://simap.eu.int/
www.ted.eur-op.eu.int
www.tendersdirect.co.uk

8

Researching value

Dr Richard Laing MRICS

Introduction

The discipline of the quantity surveyor has been transformed in recent years, and changes in the higher education syllabus mean that the modern quantity surveyor is equipped with a deep understanding of value which can impact on construction practice, project performance and the positive impact of buildings in society. Perhaps coincidentally, a series of industry- and government-supported studies have identified the need to encourage greater consideration of projects prior to construction and an equal need to consider the whole life of buildings. This chapter considers how recent research concerning value studies in this wide sense has signposted innovative and potentially lucrative routes through which the quantity surveying discipline can lead and direct modern projects from inception to disposal.

A great deal of the research which will contribute towards the forthcoming assessment of research quality in the UK Research Assessment Exercise 2008 (RAE 2008) covers material and methods which require a truly multidisciplinary approach within the design team. The appropriate unit of assessment for most surveying or built environment academic departments, Architecture and Built Environment, recognises that much of the cutting edge work being undertaken at present tends to defy traditional academic boundaries. Nevertheless, it is also true that the traditional strengths of each discipline within built environment research provides unique and valued skills.

This chapter describes streams of recent research concerning value, project assessment and innovative construction, which offer methodologies that, taken together with the traditional core skills of the quantity surveyor, provide powerful mechanisms to improve and support construction at the planning, design and occupancy stages. The surveyor's deep understanding of value, including its meaning, sources, management and relationship with external factors is vital in all cases. Case studies taken from the author's own experience are used as examples of recent research activity, and references to key work in the field are provided.

Value

The concept of value, with regard to construction and property at least, drives many decisions. Value can be taken as referring either to an all-encompassing system, or to much more focused areas within that system. It is vital that knowledge of the quantity surveyor with regard to the following is recognised and developed:

- Types of value within construction and property
- The manner in which that value can be assessed
- Understanding how value assessment can be applied.

The term value can be defined in a number of ways. The following is typical:

> **value** worth: a fair equivalent: intrinsic worth or goodness: recognition of such worth: that which renders anything useful or estimable: the degree of this quality: relative worth: high worth: esteem: efficacy: excellence: pride: precise meaning: the exact amount of a variable quantity in a particular case: (*plural*) moral principles: standards: (*verb*) to estimate the worth of: to rate at a price: to esteem: to prize

> (Chambers 20th Century Dictionary, 1983)

This definition is wide ranging in scope, and provides us with a starting point when trying to more closely define value in relation to the construction and property industries. Beginning with allusions to financial concerns, the definition proceeds to consider terms such as quality, esteem and pride. None of the definitions are more or less valid than the other, but each presents a challenge to the manager attempting to make an assessment.

Overall value comprises an amalgamation of financial, environmental, social and heritage values, preferably considered together. Rather than being a straightforward addition (of whatever variables), it is also true that certain aspects (such as internal layout, or intended end uses) may well influence more than one part of the system.

A client might demand value for money in a number of respects, and the greatest value will certainly not necessarily be delivered by the cheapest option. Value from the client's perspective is closely related to the client's other requirements from a building or property, and is therefore related to the basic original design requirements (or criteria for building choice, where the building is not new). Cost control, aesthetics, building function and control of quality must all be considered by an assessment of potential value, the likely outcome being that some or all will have to be compromised in some way. The choice of procurement method or building management strategy must ensure that the client's needs are satisfied as far as possible.

Case study: Streetscape public consultation

All streetscape projects in the public realm demand a complex response from the design team. In addition to the physical design, the reaction of users to the space will have both socioeconomic and cultural repercussions for long-term sustainability. Previous research has shown that the way people interact with the urban public environment depends largely on their perception of the physical design of that space. In addition, a high number of users could provoke increased congestion and deterioration in quality, while, naturally, too few visitors will fail to satisfy commercial investment.

It is a central role of the quantity surveyor within the modern design team to advise the client and other disciplines of how the project will influence both cost and value. This research was concerned with trying to understand how projects undertaken with the aim of improving the quality of public space could subsequently influence the economic performance of the surrounding areas. The main objective of the work was to determine those environmental investments which will sustain the population that best meets the aims of wealth creation and quality of life. The research fully recognised the importance of public participation in the design of urban public space, and sought to assist the designer through the development of reliable and reproducible methodologies. Although specific areas were studied as cases within the work, the major output was that of a method for utilising photorealistic computer models within value studies (Figure 8.1).

The public wealth, with its connotations for social, financial and environmental success, can be better assured where the public has been adequately included in the design and development processes. This work adds to a field of research concerning the use of computer models within social enquiry, which has been developing for many years. The work was also important in that it established how experimental methods taken from *environmental economics* (such as choice experiments) could be applied within the built environment, and particularly when using models traditionally used as a marketing or basic communication device within the industry (Figure 8.2).

This project provided a methodology and socioeconomic data for use by designers, and extended the applicability of choice experimentation in the built environment. The method was developed using case study sites which were selected due to the varied mix of individuals and groups which could be classed as users or stakeholders. The project clearly demonstrated that highly quantitative approaches to the measurement of *value* could be

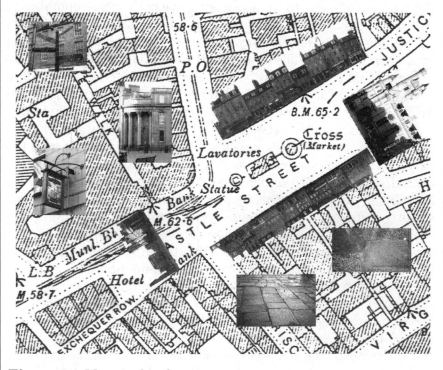

Figure 8.1 Map used in focus groups.

Figure 8.2 Still from 3D studio model.

used to record and present complex social, economic and political situations.

The project duration was 2 years, and was completed in 2001. It is true to say that computer visualisation technology has advanced significantly in the intervening time period, and that the production of similar models today would be less time-consuming. It is also worth noting that many computer-based modelling packages can either directly or indirectly store data in connection with 'objects' in the space, thus ensuring that models can serve aesthetic, cost control and project management tasks.

As mentioned, a range of assessment techniques are available to allow for value to be measured or quantified. These are illustrated in all of the accompanying case studies, with particular reference to how they can be applied in the built environment, and the manner in which recent research has helped to develop new theories to direct their use. Before we can begin to look at the practical mechanisms of value assessment in built asset management, however, it may also be useful to consider the manner in which buildings can be considered as 'assets'.

The term 'asset' can be used to define owned buildings or property, and is useful in that it regards those buildings in terms of their worth to the owner.

Definitions of 'asset' provide the following:

asset an item of property: something advantageous or well worth having

assets the entire property of all sorts belonging to a merchant or to a trading association

(Chambers 20th Century Dictionary, 1983)

The first definition is important to this discussion, as it clearly regards the *asset* as being something which is both of worth, and implies that benefits of whatever sort may be gained through that ownership. That construction will result in the production of an asset is an important step towards understanding how value assessment can be applied as part of the management process. A building, in fact, embodies a number of *assets* for the manager to consider, including the land value, building contents, space available for use and the building fabric itself. Value assessment in the built environment requires that the potential complexity of what can be offered by an asset is fully understood and that the optimum benefits are realised.

These definitions of *asset* allow us to then focus a value assessment in terms of the subject at hand. Many value assessment methodologies have been developed to consider the non-built environment, and it is essential that the approach taken in construction and property is appropriate. Value assessment frameworks are of great use where they ensure that a realistic range of possible outcomes are considered, whilst still leaving room for the decision-maker to apply his or her own expertise. The following sections discuss areas of value that should be of importance to both the design and construction processes, and also the management of built assets. Methods are also introduced through which objective or indicative assessments can be produced.

Financial value assessment

Financial value must be seen to relate to a range of both short- and long-term concerns, and an assessment should consider all stages of a building's life from construction (or the present day) onwards. Financing of construction work traditionally considered the construction period, with reference perhaps to the running costs associated with major components. Growth in the importance of longer-term financing and concerns over sustainability has meant, though, that the whole-life cost should be accounted for as early as the initial design stages. The speed at which running and maintenance costs can overtake initial construction costs means that this is most certainly in the interests of the building owner.

Life cycle costing allows for costs which may not be incurred for some time to be considered alongside present day cost (or benefits). Without this longer-term assessment, longer-term planning of finances and resources is not possible.

Where a property is purchased some time after construction, the benefits of life cycle costing are equally great, both from the point of view of the building having followed a (possibly unseen) construction process, and also regarding the building's future needs. Considering financial value from this perspective is interesting in that we can see how past interventions can heavily influence future need.

For example, the design process prior to and during construction should take account of the future maintenance needs of the finished building (e.g. life of components, fuel consumption, detail design, choice of materials). A cheaper material at the construction stage may not necessarily be cheaper in the longer term, or the relative economic benefits of a range of components may vary over time (e.g. if the anticipated life of the building is 20 or 100 years might influence the choice of the most economic component). In a similar vein, high attention to detail on site during construction might avoid unnecessary maintenance work being required at some point in the life cycle.

Likewise, where an asset is purchased during its life span, in addition to being affected by activities from design onwards, resources allocated towards its maintenance and optimisation of use will have implications in the future.

An assessment of overall financial value should clearly consider some or all of the following factors, depending on the asset itself and the purpose of the assessment:

- Construction cost
- Cost of land

- Market value of property
- Revenue generated directly by finished product
- Marketing costs
- Site maintenance costs
- Conservation considerations
- Costs associated with ensuring environmental awareness
- Gains associated with the same
- Pollution control
- The financial effect of any alterations to the existing built environment
- Demolition cost/value.

Each variable can be seen to contribute to the overall financial value system, but each will also be of importance at only certain stages in the life cycle. Construction cost, for example, will be of greater importance towards the initial stages of the building life, as it will form a major proportion of the overall cost incurred. As that proportion diminishes over time, maintenance and repair costs will increase in importance, along with the importance of potential selling value.

The inclusion of time as a variable in life cycle cost calculations both increases the extent to which the model can properly reflect reality, and also introduces additional uncertainties. The life cycle can reflect the total residual life of a building, or might represent only stages in that life (perhaps the anticipated length of the current ownership or lease). Costs or benefits which are anticipated but will not be experienced for a number of years should also be discounted to present day levels, to allow for interest rates and inflation. Clearly, predictions as to the likely costs in 20 or 30 years can be estimated, but cannot be guaranteed for accuracy. The availability of data and how those data are applied will have an impact on the final outcome and any conclusions drawn. The application of whole-life costs in practice relies on the adequate testing of any model for sensitivity to error, and the determination of whether any errors could greatly change the outcome. Although initial cost of construction (or purchase) is but one element in the list given, it will often be the only variable where the value is known. Persuading a client to choose an option where higher initial cost will give way to more attractive overall, longer-term, costs will only be possible where the life cycle calculation is seen to be reliable.

Environmental value assessment

The practice of building and architectural evaluation should include consideration of various subject areas, for example:

- Investment return
- Life cycle cost
- Aesthetic change
- Quality of detail.

The importance of aesthetic value in an overall building assessment depends on the use, occupancy and location, but should nonetheless always be considered. Although each point can be assessed to a greater or lesser extent, no one aspect can be said to represent adequately the overall value change. Indeed, it is suggested by Ilozor *et al.* (1997) that a range of factors: economy, efficiency, durability and aesthetics (or financial, environment and heritage) be considered equally, and that an overemphasis on any one part would jeopardise the consideration of others.

The term 'environmental' can refer to a wide subject area and this section explores definitions and limitations of those variables in greater depth. As concepts of environment vary or develop, approaches appropriate for assessment will also change. The potential importance of stone cleaning as an instrument of environmental change must be recognised. A major part of the environmental value subsystem should be taken to include the value placed on buildings by those using and occupying them, as well as any indirect value obtained by the wider population. For example, with prominent public buildings, it is likely that non-use value would account for a significantly larger proportion of overall value than where less prominent buildings are to be considered. For example, even where an individual does not plan to visit a building themselves, it is possible that in certain cases the value placed on that building by them might be considerable.

The term 'environmental value' is commonly used to cover value:

- derived by individuals through use of, and contact with, an asset
- derived by individuals neither using nor visiting the asset
- due to any perceived social benefits/costs.

Numerous methods have been developed to increase objectivity in the assessment of environmental value, most of which were initially developed in relation to the natural (i.e. non-built) environment. Prior to the beginning of any new construction work, or major alterations to existing buildings, the design and planning processes are required to adequately consider environmental value. In practice, this includes a range of factors:

- The visual appeal of the public façade
- Consideration of surrounding buildings
- Implications for the local community
- Potential use of the building
- Public opinion.

Methods commonly used to objectively appraise environmental value either attempt to directly measure the environmental value of a project, or uncover links indirectly represented in existing markets. For example:

direct the social effects of a proposed project can be 'measured' by directly asking the public (or another group) for both their opinions and evaluation of the proposals. For example, previous research has indicated that the public would be willing to contribute towards funds to ensure the continued protection of buildings of historic interest. Even if no actual payment is received, a 'measurement' of sorts is established.

indirect effects of the proposed work within some other market can be investigated. For example, the link between property markets and air quality has been established. Therefore, proposals for an asset development which might result in air pollution could reasonably be said to have a potential for negative environmental value.

Preservation of the environment, in both the built and natural form, may be facilitated through the promotion of sustainable development. Pearce and Markandya (1989) define total economic value as being the total user benefit plus the total intrinsic benefits of an environmental good. However, the assessment of total economic value also tends to rely on the knowledge of the respondent group. How can we estimate the value of something we do not understand? Care must be taken to understand how environmental values can be related with other areas of value, ensuring that no parts of the value system are overlooked due to simplification of the subject matter.

When defining user benefits, these are normally understood to comprise both consumptive and non-consumptive aspects. In terms of the built environment, these would be in the form of financial return (consumptive) as well as, for example, the enjoyment gained from the visual appeal (non-consumptive) of the built landscape. Many of the benefits that may be derived from stone cleaning, in the short term at least, could be described and categorised as being user benefits.

The following case study took as its starting point the notion that certain types of open space, which most people would regard as being vital to the continued quality of our lives and communities, existed outside the property markets. A problem was therefore evident in that such spaces often had to compete against revenue generating spaces or activities for public funds, yet it was difficult to demonstrate how the spaces themselves could justify continued expenditure. It is perhaps worth noting that national governments throughout Europe have recently launched a series of policies and strategies to encourage the use of shared spaces for healthy living and sporting activities. Therefore, the *value* of such spaces is beginning to become enshrined through political rather than economic support.

Case study: Public open space

Urban greenspace makes an essential contribution to quality of life. It provides a recreational resource, a peaceful retreat from the city, an attractive backdrop to built development, safe and exciting play areas for children and a reserve for urban wildlife. Many municipal authorities do not receive any significant income from greenspace, but must budget for its maintenance along with other municipal responsibilities such as education or roads. Consequently, many greenspace areas are either neglected or are sustained in a form that risks becoming less relevant to modern life styles.

Working as part of a Europe-wide consortium, the 'Greenspace' project team undertook an analysis of the use and values attached to different greenspace types; appraised the application of qualitative techniques and quantitative methodologies, in particular the combined use of choice experimentation and visualisation; and analysed the use of public participation to explore greenspace needs. The needs of decision-makers were central to the research, which aimed to support key decisions faced by authorities across the continent.

Also of interest was the impact which geography and culture had on the various methods used across Europe. For example, research teams based in countries where public participation was arguably well positioned and established within public policy (e.g. Germany, The Netherlands) were able to apply extremely arduous studies of greenspace users, often involving large amounts of time and effort on the part of respondents. Conversely, the research team based in Spain required to use methods which required less direct involvement of respondents, but which could still elicit the necessary preference data. An important lesson was that although the theoretical basis for many research approaches can be applied across many studies, the practical application will usually require adjustments to recognise the context within which studies are being undertaken (e.g. data availability, format of data, culture of respondents or organisation).

The innovative aspects of the Greenspace study really concerned the assessment of how people value the greenspace to which they already have access, and also the development of approaches whereby people can inform designers and decision-makers regarding the provision of new space. Although this study concentrated to some extent on the development of visualisation and other computer-based tools to help design teams and local authorities, there was also considerable work considering the use of focus groups (of greenspace users) to identify key social, economic and design factors. The research teams from the UK included members with expertise in quantity surveying, estate management and architecture, as well as computer modelling and economics. This mix of disciplines meant that the research was able to employ a varied range of methods, and that the outputs were presented in such a way that potential professional use of the data was considered from the outset.

Results from the study will help to guide the future planning of greenspace provision, in the UK and abroad, and develop methods to assist planners, designers and public in the pursuit of greater levels of public participation. Furthermore, the arrangement of European level research is such that it has been possible to form a continent-wide organisation of research teams entitled the 'Greencluster'. A central aim of the cluster is to promote the strength or undertaking such research when coupled with practice, and as such has helped to influence policy at the national level. This has included, within the UK, participation in a number of national conferences, and contributions to events and studies undertaken by Greenspace Scotland, the Urban Parks Forum and the Scottish Executive.

Further details of the research can be found via www.ucd.ie/greensp/ and www.greencluster.org/

This case study is important in that it demonstrates the manner in which methods taken from traditional building economics and planning can be used in connection with disciplines as diverse as botany and environmental monitoring to produce highly effective and practical tools for immediate application. Again though, the focus throughout all the studies covered under the 'Greencluster' was to understand how value, and those factors influencing value, can be influenced through a prudent understanding of user needs and environmental and political demands.

From the perspective of asset management, these methods to support value assessment are of interest (particularly if previous work has explored similar projects to those being considered), but might seem time-consuming or unmanageable to the non-expert. The importance of the concepts being explored, however, cannot be overstated. A major part of the value system associated with any building will concern non-financial variables, and that must not be overlooked by the manager.

Heritage value assessment

Case study: Masonry conservation and whole-life costs

This case study concerns a stream of building conservation research which has been undertaken by various multidisciplinary teams since the 1980s. In particular, it focuses on the use of value assessment techniques to assess the impact which the practice of uncontrolled stone cleaning, often using technically unsound methods, had on the value of those buildings treated.

Across Europe, stone cleaning has been applied widely to many buildings over a period of more than three decades, producing a varied range of results. This research was concerned with the development of reliable methodologies which can be employed as part of a decision-making process, to help ensure that future stone cleaning takes full account of the implications for overall value (overall value being conceived as the aggregate of financial, environmental and heritage values). Latterly, the work came to concentrate on the long-term effects of masonry cleaning and conservation, including specific issues for the supply chain.

The approach to overall assessment emanating from this research structured a series of assessments, ensuring that gains in the short term cannot over-ride potential losses over the remaining life cycle. An ultimate aim of all stone cleaning is to enhance the built environment in some respects. The aim of this value assessment was to ensure that cleaning is completed only where an overall gain or benefit in value is attainable.

An important study, funded by Historic Scotland, considered the long-term consequences of stone cleaning. That study demonstrated how methods from surveying, geology, chemistry and geography could be used together to provide clear methods for use by designers, managers and building owners. The work includes a detailed assessment of the life cycle costs of cleaning, stone repair and stone replacement (Young *et al.*, 2003). It is interesting to note that a key assumption of the model is that stone will actually be available to undertake repairs, whilst the situation in much of the UK is that many quarries commonly used in the past have been mothballed for many years. The consequence of this is that very significant engineering works may be required to enable the extraction of even small amounts of stone.

Overall value assessment

The consideration of value in relation to the built landscape demands that a wide view be taken, in that the effects of any change will influence financial, social and cultural aspects of that environment. A value assessment concentrating on one particular area of value could be accused of considering the views of a minority, at the expense of society as a whole. Legislation and guidelines concerning protection of the built environment present approaches to the planned development of the built environment that will protect the longer-term needs of society, and ensure a sustained protection of the environment in which the greater part of western society lives.

Whatever methods are employed in the assessment of value, it must be ensured that a wide and inclusive approach is taken, both conceptually and practically.

Value-driven management techniques

The preceding discussion has indicated how value can be used to encompass many of the important factors to be considered in relation to the built environment (e.g. finances, client needs, aesthetics, social value). The manager of built assets must attempt to satisfy the needs of each part of the value system, thus ensuring that the best possible outcome is achieved.

Maintenance of an asset is essential in the sense that deterioration of a building (either in use or otherwise) is inevitable and must be addressed. Most aspects of value will deteriorate where the fabric of the asset deteriorates, and a reaction is necessary. The often reactive nature of maintenance work, however, actually hinders both the overall value system and the ability to plan due to the uncertainty of when further maintenance will be required. Planned preventative maintenance strategies seek to avoid uncertain and unexpected future costs by anticipating where faults are going to occur. Information regarding anticipated life span, life cycle data and the client needs can provide the basis for an initial maintenance plan.

Case study: Urban redevelopment and renewal

This research concerns a number of key issues which are of central importance to understanding public participation in the context of urban design.

It is thought to be essential that underused areas are developed in such a way that better physical and other connections can be generated across urban centres. This research is considering and modelling potential solutions to the issue of disconnection, and will undertake to gather and report a detailed study of associated public perceptions, attitudes and values.

Cities throughout the UK and Europe often rely on master planning to direct the development of their city centre growth, yet such plans by necessity rarely include detailed proposals for exactly what should be constructed. This research takes such master planning statements as an important starting point, from which community- and user-led designs can emerge. Similarly, city centres rarely exist as cohesive entities, but rather tend to develop naturally into a collection of associated but distinct neighbourhoods or districts. It is similarly important that physical, social and aesthetic connection between such spaces are not allowed to disappear or become diminished, due to changing uses or due to well-meaning developments which in hindsight reduce the usability or accessibility of the city (Figure 8.3).

The importance of major planning and design proposals for life in our cities cannot be underestimated, and the potential value of establishing an accessible and integrated city as a result would be significant for residents and visitors alike. To a great extent, though, the success or failure of design proposals will depend on their being embraced by users, thus providing an economic foundation and helping to support a self-sustaining longer-term development of the city.

This research has employed computer visualisation and cultural perception studies to gather vital data pertinent to specific areas within Aberdeen. In the short term, this data could be used

Figure 8.3 Detail taken from a project study model.

to develop, evaluate and refine specific design solutions for the area. It is of utmost importance to gain a clear understanding of peoples' perceptions and understanding of the city centre as a social and physical landscape (http://rgusurvey.org.uk/aberdeen/).

Although drawn up initially either immediately after construction or after the asset is purchased, any maintenance plan should evolve over time to incorporate new data as they become available. Manufacturer's information concerning likely life spans is based on averages which may be different to those experienced by the client. Through a process of data recording and communication, a planned maintenance programme can be tailored to the needs of the individual facility, thus providing a truly pro-active system and allowing for an accurate life cycle costing.

Case study: Building maintenance whole-life costs

Recently, there has been a fundamental desire to adopt a whole-life attitude regarding the design and management of buildings because of the dramatic shift in the balance between the initial capital cost and the running costs of buildings towards a substantial increase in the running costs.

Perhaps one of the challenging obstacles facing this desire is the fact that the design or component selection decisions can often be taken based on factors other than cost criteria. This is especially true in the complex environment of healthcare buildings, in which, for example, the desire to reduce variation for economic reasons has to be balanced against a wide variety of specialist uses and a large number of user groups with widely differing needs.

This project, funded through NHS Estates, aims to develop an integrated system for the optimal selection of hospital finishes. Essential requirements for optimal hospital design are to be included within the model, with emphasis on their consequential implication for the selection of finishes. This includes a wide range of complex and holistic design requirements that have a fundamental influence on the value and quality of life in hospital environments including, for example, planning, legislative, space, psychology, aesthetic, risk, economic, as well as many others.

Initially, design requirements of various types of finishes, including flooring, walling, ceiling, doors, windows and fitting/furnishings are briefly discussed before narrowing the focus of the research to one finish type. Then, a staged process for the identification of selection criteria is proposed.

The proposed framework employs a database management system as a data repository. A number of applications are provided to interact with the data repository through an interactive user interface, the applications aiming to generate feasible alternatives and criteria for a given application, and the identification of the optimal alternative. This includes a whole-life costing application, an evaluation tool to rate various options in respect of qualitative criteria, and an application to elicit weights of importance of various criteria from various stakeholders.

This work is timely in the sense that there has been a growing pressure on healthcare organisations to help ensure that the buildings in which they operate are safe, and carry no additional risk of infection. The model itself will be required to convey a complex range of information to the end user, so that it can genuinely assist in the selection of finishes. The past few years have also clearly demonstrated that the design of healthcare environments requires designers to properly consider the needs of patients and staff, whilst also recognising that spaces require to be flexible. Therefore, it may not be possible in all cases to identify at the point of construction how a space will be used throughout the life span. The ability of modern whole-life cost systems to cope with such uncertainties and present a range of possible outcomes to the designer will become a vital part of the quantity surveyors skill base in years to come. Therein, the use of cost models will almost inevitably better meet the needs of the design team and allow the team (and client) to more effectively balance costs with functional and aesthetic requirements.

The model also represents a topic of great importance to the future of the quantity surveyor, in that it demonstrates how techniques originally developed to determine long-term financial costs are now being used to predict the qualitative as well as quantitative implications of design decision. In this instance, it is clear that core skills and knowledge surveyors and cost engineers have a vital part to play in the complex arena of healthcare design.

Optimisation of most assets involves a system of data collection, analysis and feedback. Data are collected regarding the performance of an asset or group of assets owned by the client (e.g. annual running costs percentage of time out of order). From these data, the best performance recorded – the 'optimum' – can be established. A suitable range outside this where performance would be deemed acceptable can be decided based upon client need, cost incurred or past experience, and defective assets identified. Where assets of whatever type are identified in this way of having less than acceptable performance, a programme of replacement can be introduced.

The optimisation process can be applied to any part of the built asset be it part of the building fabric or otherwise. If operated alongside a preventative maintenance strategy, the potential benefits are evident.

Discussion

The concept of value in the built environment has always played a major role in decision-making, but has not always been properly identified within the design, construction and management processes followed. It is also true that whilst wider concepts of value are widely accepted within society, that these have not always been properly included within cost and value studies in construction and planning.

Available management techniques allow for a detailed financial valuation whilst helping to maintain the environmental values described earlier. The value system associated with the built environment provides a framework for decision-making which has the potential to greatly strengthen and enhance the future role of the quantity surveyor.

Bibliography

Al-Kodmany, K. (1999). Using visualization techniques for enhancing public participation in planning and design: Process, implementation, and evaluation, *Landscape and Urban Planning*, **45**, 37–45.

Department for Environment, Food and Rural Affairs (DEFRA) (2003). Sustainable development – the UK Government's approach, *Sustainable Development*, updated 20 August 2004, viewed 1 November 2005, http://www.sustainable-development.gov.uk/index.htm

Ilozor, B.D., Oluwoye, J.O. and MacLennan, H. (1997). The concept of aesthetic values in the selection of building project alternatives. *Journal of Real Estate and Construction*, **7**, 53–69.
Nasar, J.L. (1998). *The Evaluative Image of the City*, Sage Publications, London.
Pearce, D.W. and Markandya, A. (1989). *Environmental Policy Benefits: Monetary Valuation*, OECD, Paris.
Urban Task Force (1999). *Towards an Urban Renaissance*. London, E&FN Spon.
Young, M.E., Ball, J., Laing, R., Cordiner, P. and Hulls, J. (2003). *The Consequences of Past Stone Cleaning Intervention on Future Policy and Resources*. Historic Scotland, Edinburgh.

Websites

CABESpace http://www.cabespace.org.uk/index.html
CABESpace was established to influence open space strategy and activity in England. During 2003, the organisation commissioned its first raft of research projects, final reports from which are available for download.
ODPM Sustainable Communities http://www.odpm.gov.uk/stellent/ groups/odpm_urbanpolicy/documents/sectionhomepage/odpm_urban-policy_page.hcsp
PAN65: Planning and Open Space (2003) http://www.scotland.gov.uk/ library5/planning/pan65-00.asp
TPL, The economic benefits of open space (http://www.tpl.org/ tier3_cd.cfm?content_item_id=1195&folder_id=727)

Appendix

The following pages show examples of notices to be posted by public sector contracting authorities in order to comply with EU Public Procurement Directives, as previously discussed in Chapter 7:

(1) A prior information notice (PIN); (2) An invitation to tender notice; (3) A contract award notice (CAN) (Source: Tenders Direct).

These notices have been published in the Official Journal and all relate to the same contract, i.e. the provision of project management and quantity services for University College London. It should be noted that had the purchase authority been other than UK based, then the majority of the notices would have been published in the authority's native language.

Figure A.1 shows a Prior Information Notice (PIN) that is published either at the beginning of the year or more usually in the case of construction contracts when the project is first planned. A PIN notice is designed to provide contractors and professional services companies with advance information that a project is about to be put out to tender so that they can begin to prepare their response.

Figure A.2 shows an Invitation to Tender Notice which provides information on the services required, as well as contact details for the purchasing authority and the date by which potential suppliers must have responded.

Figure A.3 illustrates a Contract Award Notice which is required after the contract has been awarded and it lists the name and address of the successful supplier.

Prior-information procedure

Title:	UK-London: project management and quantity surveying services
Purchase Authority:	UNIVERSITY COLLEGE LONDON
Journal Ref:	123386-1999
Published on:	02-Sep-1999
Deadline:	This notice expired on 19/08/00.
Contract Type:	This is a service contract.
Country:	United Kingdom
Notice Type:	Prior-information procedure
Regulations:	This document is regulated by the European Services Directive 92/50/EEC.

Tender Details

1. Awarding authority: University College London, Estates and Facilities
Division, Gower Street, UK-London WC1E 6BT.
Tel. (01 71) 391 12 41. Telex 28722 UCPHYS-G. Telegraphic address:
University College London. Facsimile (01 71) 813 05 24.
2. Intended total procurement (Annex I A): CPV: 74142100, 74232400.
CPC reference No 867.
Project management and quantity surveying services.
University College London will be seeking expressions of interest from consortia for the provision of the above services for a contract to rebuild and refurbish a building in Huntley Street, UK-London. The proposed building is to be 5 storeys, plus basement and sub-basement.
The value of the building contract will be in the order of 25 750 000 EUR (17 000 000 GBP) and service providers'' contracts related accordingly.
3.
4. Other information: The estimated date for the awarding of these
contracts will be 12/1999.
Additional information may be obtained from the address in 1.
5. Notice postmarked: 19. 8. 1999.
6. Notice received on: 23. 8. 1999.
7.

Figure A.1 Prior Information Notice.

Invitation to Tender Notice

Title:	UK-London: project management and quantity surveying services
Purchase Authority:	UNIVERSITY COLLEGE LONDON
Journal Ref:	127831-1999
Published on:	15-Sep-1999
Deadline:	This notice expired on 15/10/99. Click here to view details of the award.
Contract Type:	This is a service contract.
Country:	United Kingdom
Notice Type:	Invitation to Tender Notice - Restricted Procedure
Regulations:	This document is regulated by the European Services Directive 92/50/EEC.

Tender Details

1. Awarding authority: University College London, Estates and Facilities
Division, Gower Street, UK-London WC1E 6BT.
Tel. (01 71) 391 12 41. Telex 28722 UCPHYS-G. Telegraphic address:
University College London. Facsimile (01 71) 813 05 24.
2. Category of service and description, CPC reference number, quantity,
options: CPV: 74142100, 74232400.
Category 12, CPC reference No 867.
Project management and quantity surveying services from consortia for a
contract to part rebuild and part refurbish a building in
Huntley Street, UK-London. This building was built in the
early 1900s. It is proposed to part rebuild on the site and
part refurbish the existing building to form research
laboratories, associated offices, ancillary areas and some
teaching areas. The building will be 5-storeys plus basement
and sub-basement.
The value of the building contract will be in the order of
25 750 000 ECU (17 000 000 GBP). The service providers''
commission will be from 2-3 years, and contracts will be
related in value accordingly. It is envisaged that a maximum
of 5 consortia will be shortlisted and invited
to tender. The consortia should agree a lead consultant to
respond to this notice.
3. Delivery to: UCL, UK-London.
4. a) Reserved for a particular profession: UCL require that
organizations undertaking building consultancy work employ
on that work
staff having the relevant professional experience and
competence for that job.
4. b) Law, regulation or administrative provision: The
requirement in 4
(a) is an administrative provision of the university.
4. c) Obligation to mention the names and qualification of
personnel:
The names and professional experience of the staff to be
responsible for the execution of the services will be sought
from those invited to tender.
5. Division into lots: The service provider can only tender
for the whole of the services. Where different disciplines
form a consortium, the lead consultant should respond to the

Figure A.2 Invitation to Tender Notice.

```
notice.
6. Number of service providers which will be invited to
tender: A maximum of 5 consortia will be shortlisted and
invited to tender.
7. Variants: Variants will not be accepted.
8. Time limits for completion or duration of the contract,
for starting
or providing the service: 2-3 years.
9. Legal form in case of group bidders: Joint and several
liability.
10. a)
10. b) Deadline for receipt of applications: 15. 10. 1999.
10. c) Address: Mr T. Edwards, UCL, Estates and Facilities
Division, 1-19 Torrington Place, UK-London WC1E 6BT.
10. d) Language(s): English.
11. Final date for the dispatch of invitations to tender:
30. 10. 1999:
this date may change subject to level of initial response.
12. Deposits and guarantees: No deposits or guarantees
required.
13. Qualifications: Proof of the service provider''s
financial standing
must be furnished by:
appropriate statements by bankers;
submission of latest audited annual report and balance
sheet;
statement of turnover, and turnover in the previous 3
financial years;
a list of relevant principal contracts undertaken in the
past 5 years,
details to include values, dates, description and clients;
description and evidence of quality-control procedures;
indication of membership of technical bodies of those staff
responsible
for the provision of the service;
details of 2 current clients, their names and addresses, who
are willing to supply references;
evidence of professional-indemnity insurance.
14. Award criteria: Relevant experience of tenderers and
professional competence, quality and technical merit of
tenders. The economically most advantageous tender.
15. Other information: If invited, tender bids will be in
sterling.
16.
17. Notice postmarked: 25. 8. 1999.
18. Notice received on: 6. 9. 1999.
19.
```

Figure A.2 Continued

Contract Award Notice

Title:	UK-London: project management and quantity surveying services
Purchase Authority:	UNIVERSITY COLLEGE LONDON
Journal Ref:	51239-2000
Published on:	21-Apr-2000
Deadline:	N/a.
Contract Type:	This is a service contract.
Country:	United Kingdom
Notice Type:	Contract Award Notice
Regulations:	This document is regulated by the European Services Directive 92/ 50/ EEC.

Tender Details

```
1. Awarding authority: University College London, Estates &
Facilities
Division, Gower Street, UK-London WC1E 6BT. Tel.: (020) 76
79 12 41.
Telex: 28722 UCPHYS-G. Telegraph: University College London.
Fax: (020)
76 79 05 24.
2. Award procedure chosen, justification (Article 11(3)):
Restricted.
3. Category of service and description, CPC reference
number, quantity:
CPV: 74142100, 74232400.
Category 12; CPC reference No 867.
Project management and quantity surveying services.
4. Date of award: 12/2000.
5. Award criteria: Selection based on weighted scores for
the following:
competence, quality and technical merit of tenderer,
economically most advantageous tender.
6. Tenders received: 41.
7. Service provider(s): Mace.
8.
9.
10.
11.
12. Contract notice published on: 15.9.1999.
1999/S 179-127831.
13. Notice postmarked: 7.4.2000.
14. Notice received on: 11.4.2000.
15.
```

Figure A.3 Contract Award Notice.

Index